How To Make DISPOSABLE SILENCERS

A Complete Guide

Desert Publications
El Dorado, AR 71731-1751 U. S. A.

How to Make Disposable Silencers

Published by Desert Publications
215 S. Washington Ave.
El Dorado, AR 71730 U.S.A.
1-870-862-3811
info@deltapress.com

Printed in U.S.A
ISBN: 978-0-87947-163-7
10 9 8 7 6 5 4

Desert Publication is a division of
The DELTA GROUP, Ltd.

Direct all inquiries & orders to the above address.

WARNING NOTICE

The construction or possession of many of these devices would be in violation of various Federal, State, and Local laws. Please contact your local B.A.T.F. office for any information you may require, BEFORE attempting the construction of any of these devices. Severe penalties are prescribed for violators of these laws!

The publisher assumes no responsibility for the use or misuse of any information contained in this book.

These articles are presented for academic study only!

TABLE OF CONTENTS

INTRODUCTION

One of the problems man has had to contend with throughout the history of firearms, has been noise. Prior to discharging an unsilenced weapon, some shooters have the tendency to flinch in anticipation of the muzzle blast, thus affecting accuracy. After discharge, a bigger problem is the period of temporary deafness the shooter experiences.

Gun noise also attracts attention, and for individuals or small groups of combatants, facing a superior force, this could mean detection and disaster.

Gun noise can be categorized into three major categories.

1. MECHANICAL NOISE

The noise the weapon's mechanism makes in order to fire a round.

This noise would be least noticeable in a bolt action weapon, due to its minimum of moving parts, and most noticeable in an open bolt submachine gun type weapon or semi-auto weapon. In real practice, however, all three weapon systems have been effectively silenced, examples being the MAC 10/11 submachine guns and the RUGER MK1 and HIGH STANDARD .22 caliber semi-auto pistols.

2. SONIC CRACK

The noise a bullet makes when traveling over 1100 fps (feet per second).

When a bullet travels at super sonic velocities there will be a sonic crack much like a jet's sonic boom. Little can be done to lower the velocity of bullets which travel at a much higher level than 1100 fps. Extensive modifications to the weapon's barrel system would be required. One modification could be porting the barrel to release propellant gas before the bullet exits the barrel to reduce velocity. This is undesireable in certain calibers as they require a high velocity in order to be effective. Case in point is the .223 Remington (5.56mm).

Other calibers which are just slightly above the 1100 fps level, as is the case with the 9mm luger, can be properly lowered to subsonic levels by handloading with lighter powder charges or using heavier bullets. If using .22 long rifle caliber ammunition, try to avoid hyper velocity rounds as some exceed super sonic levels.

3. MUZZLE BLAST

The noise generated by the escape of the hot propellant gas.

Of all the noises associated with firearms, the most significant is the muzzle blast, which is caused by escaping propellant gas upon bullet exit from the barrel.

The objective of most silencer designs is to reduce the muzzle blast by preventing the sudden release of the hot propellant gas long enough to reduce its pressure and lower its temperature.

In constructing any silencer, there are a few points that should be considered first, such as the purpose for which the device will be used and the degree of silence desired. It should be remembered that complete silence has rarely been achieved. For maximum effectiveness, a caliber should be chosen that is below super sonic velocity and that does not produce so great an amount of escaping gas as to make the silencer too large and bulky, thus impractical.

Whenever possible, avoid revolvers as they usually can not be effectively silenced due to the escape of gas between the barrel and chambers.

CONSTRUCTION SECTION

The following designs were mainly selected for their ease of construction, material availability, low cost, and most important of all, effectiveness.

The construction of these silencers were designed to use a minimum of materials, labor and tools. Materials we researched to make the silencer as effective as possible, yet easily accessible at any supermarket or hardware store. These designs, though they may appear crude, where found to be the most basic and effective. All conform to the fundamentals of effective firearm silencer principles.

One important point to bear in mind is that these silencers, although very effective, are "disposable" silencers, and will lose their effectiveness after a few shots.

One of the main features of these silencers, besides their low cost, is their legality. Federal, State and Local laws prohibit the possesion of any firearm silencing device without a special license. You must first apply for the license, then if approved, pay a special tax of $200.00 and then proceed to assemble your silencer. Thus the design of these silencers are such that owning all parts by themselves (DISASSEMBLED!) is perfectly legal. ALWAYS remember that any ASSEMBLED device, whether it be improvised or professionally made, to reduce the sound level of a firearm is illegal without this special license. In some states you are not given a choice, as all sound suppressors are prohibited to private citizens. Check with your local authorities BEFORE construction.

Although most of the silencers presented in this book can be installed to practically any weapon, to achieve maximum effectiveness, the author recommends the use of a .22 long rifle caliber weapon.

Experimentation is a key element in developing an effective silencer. Different diameters, lengths, and shapes may have a great effect on the effectiveness of any given design.

(NOTE: Dimensions in this book are given as references only. Depending on the caliber of your weapon, you may require a different length or diameter or both.)

SCREEN SILENCER

Here is one that is very effective for .22 caliber weapons.

2 pieces of screen, 6 inch and 8 inch wide by several feet long, some tape and an ordinary lead pencil are all the materials required.

Start with the strip of screen wire six inches wide and several feet long and begin wrapping one end around the pencil. Continue to wrap the screen wire around the pencil until the roll is the same diameter as the outside diameter of the barrel on your weapon.

Use two pieces of tape to prevent the screen wire roll from unwrapping.

Now switch to the strip of screen wire about 8 inch wide. (NOTE: It is suggested that the front end of the barrel be taped to prevent it from being scratched by the screen.) Keeping one edge flush with the front of your original roll, let the other edge extend back two inches around the barrel. Each time the screen wire is wrapped over the front sight, cut a small notch in the wire, so that the sight extends through it, and press that layer down tightly against the layer beneath it. Continue wrapping four or five turns until the inner cylinder of screen wire is firmly held to the gun barrel. Cut off any excess length of the screen wire strip.

Finally, wrap the outer part of the roll with two layers of ordinary electrical tape or cloth adhesive tape, letting the tape extend back an additional two inches on the bare gun barrel to hold the "silencer" in place. Remove the pencil before firing.

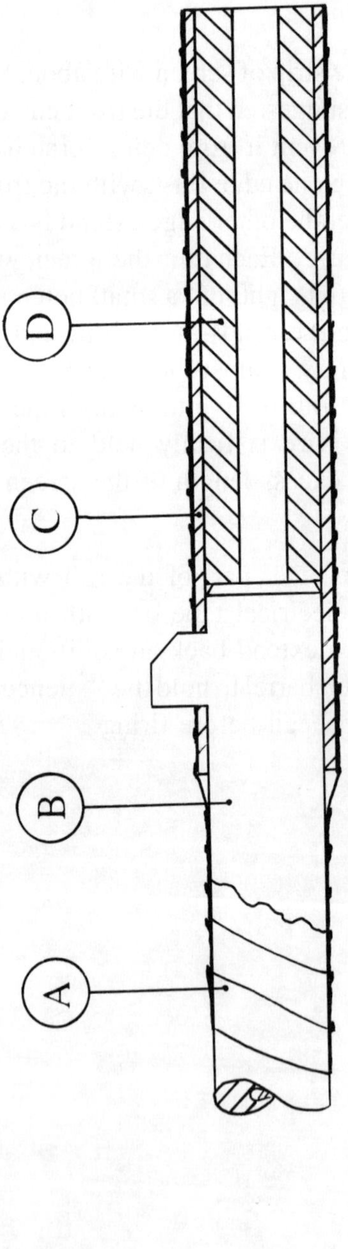

A) TAPE B) BARREL C) OUTER WINDOW SCREEN ROLL D) INNER SCREEN ROLL

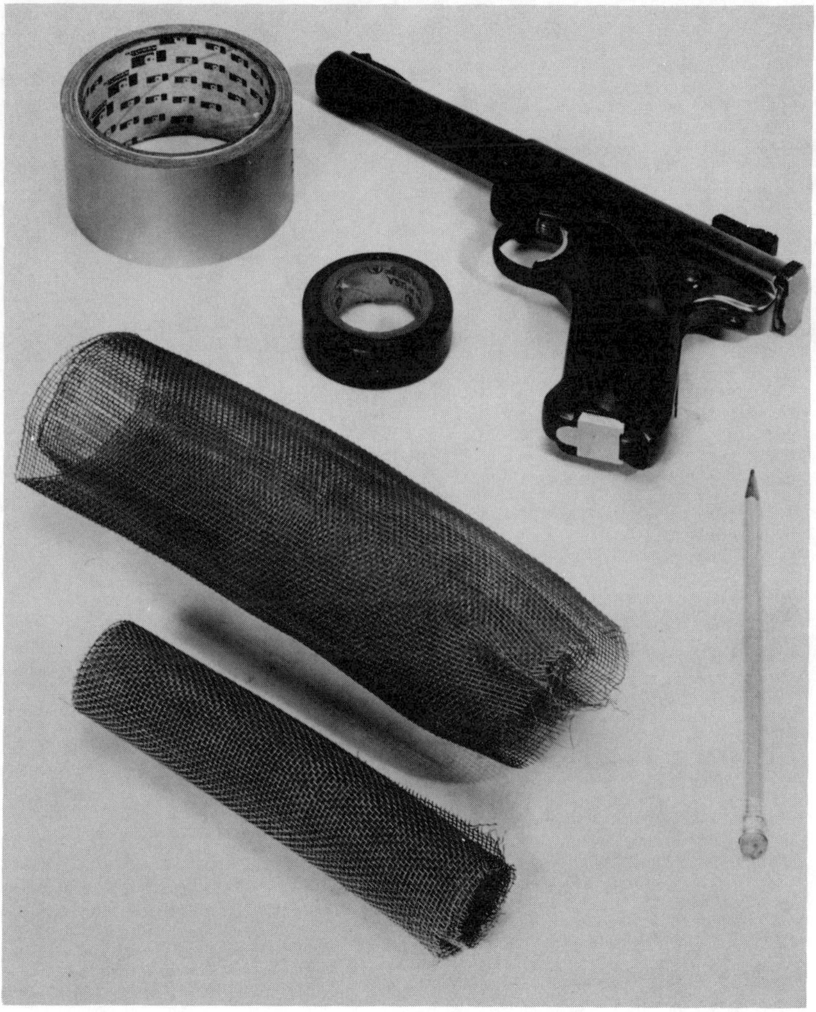

Materials Required: A) Cloth adhesive tape B) Electrical tape C) Pencil or cleaning rod D) Window screen.

For easier construction, front sight can be removed. If removing sight is impractical, refer to text for instructions.

Tape barrel to prevent it from being scratched by screen.

Measure screen. Two widths will be needed. Size will depend on caliber of weapon used. For .22 caliber long rifle, a 6″ and 8″ will work fine.

Start by rolling screen around pencil.

Finish inner roll of screen. Hold together with tape.

Inner screen roll in place with outer screen ready to be wrapped.

Wrap outer screen four or five turns around inner roll. Keep front edge flush with the front of original roll.

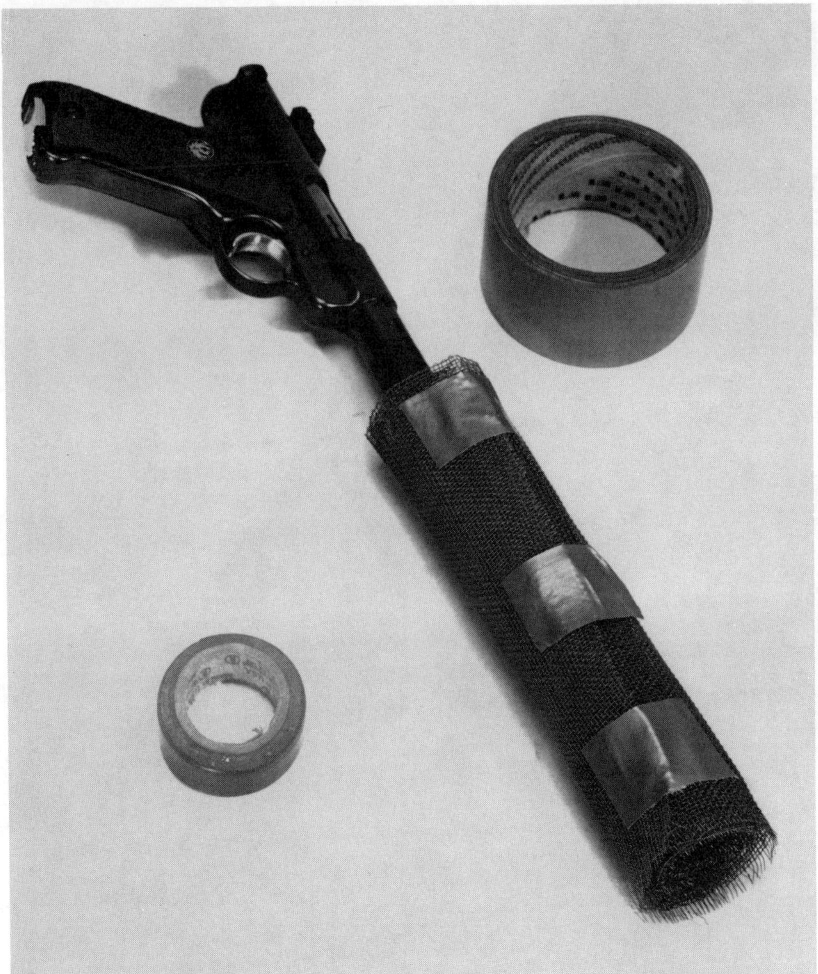

Once outer screen has been completely wrapped, hold in place by using strips of tape.

Wrap outer part of the screen with two layers of tape.

Completed unit. If desired, for maximum effectiveness, front end of unit could also be taped.

SOFT DRINK BOTTLE SILENCER

The following design is one of the simplest to manufacture, least expensive and yet, one of the most effective of all disposable silencers.

All that is required for construction is an empty 1 or 2 liter plastic soft drink bottle, a 1 or 1½ inch hose clamp and some tape.

The bottle chosen for this purpose must have a reinforcing ring on the bottom portion, this is necessary to prevent the bottle from splitting upon discharge. There are several brands of soft drink that have this reinforcing ring, so get one that you will enjoy drinking.

This unit is only effective for a few shots, as the noise level will increase with each successive shot, due to the enlarging exit hole at the bottom of the bottle.

Begin construction by measuring the inside diameter of the bottle neck with the diameter of the barrel. If the diameter of the barrel is smaller, bring it up to the required size by using electrical tape. (NOTE: With some weapons you will need to make a slot on the bottle neck for the front sight. Make it a tight fit!).

Next, slip the 1 or 1½ inch hose clamp over the bottle neck and install bottle to weapon. Once the bottle is in place, position the hose clamp over the tape and tighten clamp. (Make sure there is a good seal between the tape and inside of bottle neck.) Check the front sight slot for any openings, and use tape, if necessary, to seal any leaks.

If desired, use some paint to match the "silencer" to your weapon's finish.

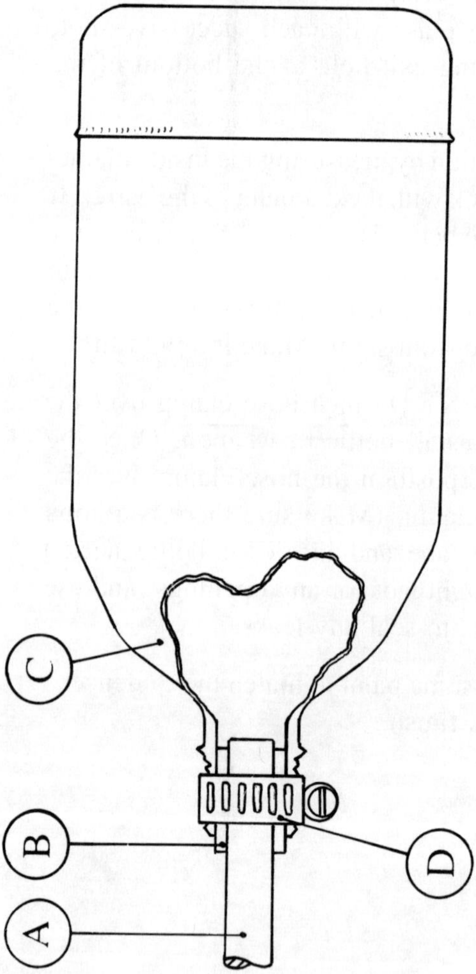

A) BARREL B) TAPE BUSHING C) SOFT DRINK BOTTLE D) HOSE CLAMP.

Materials needed for suppressor: A) One, or two liter soft drink bottle B) Electrical tape C) One inch hose clamp.

Constructing barrel bushing from tape.

Close-up of barrel bushing.

Install hose clamp to bottle neck.

Install bottle to weapon by tightening hose clamp.

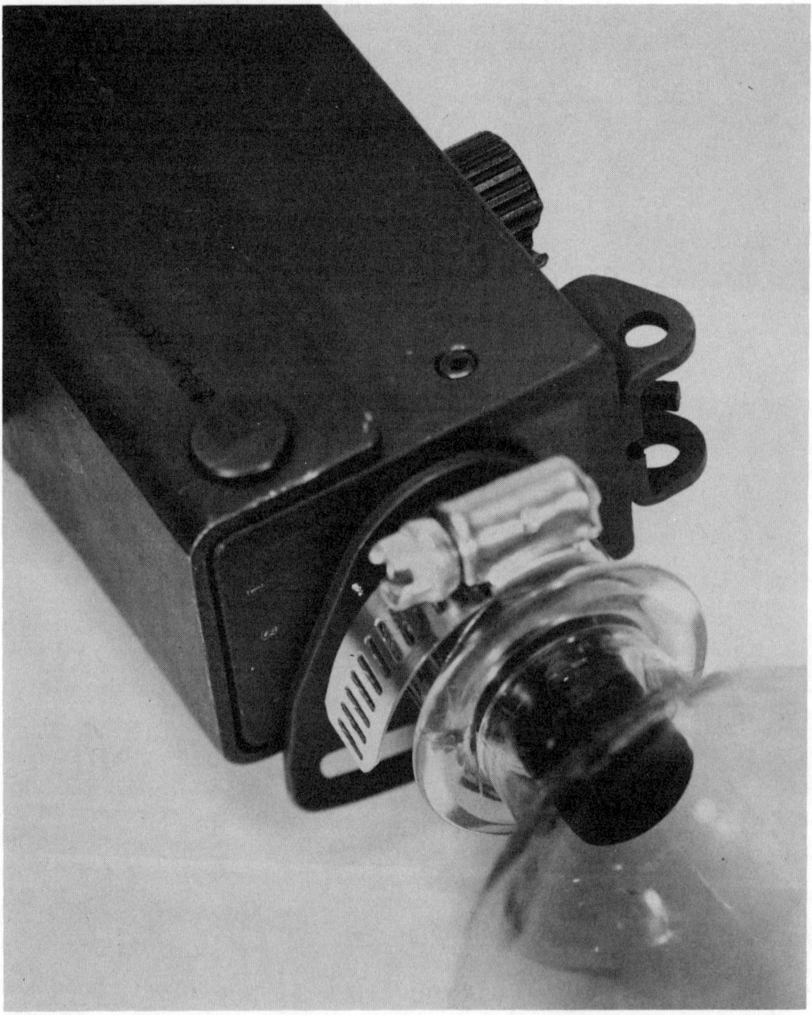

Close-up of suppressor mounted on barrel.

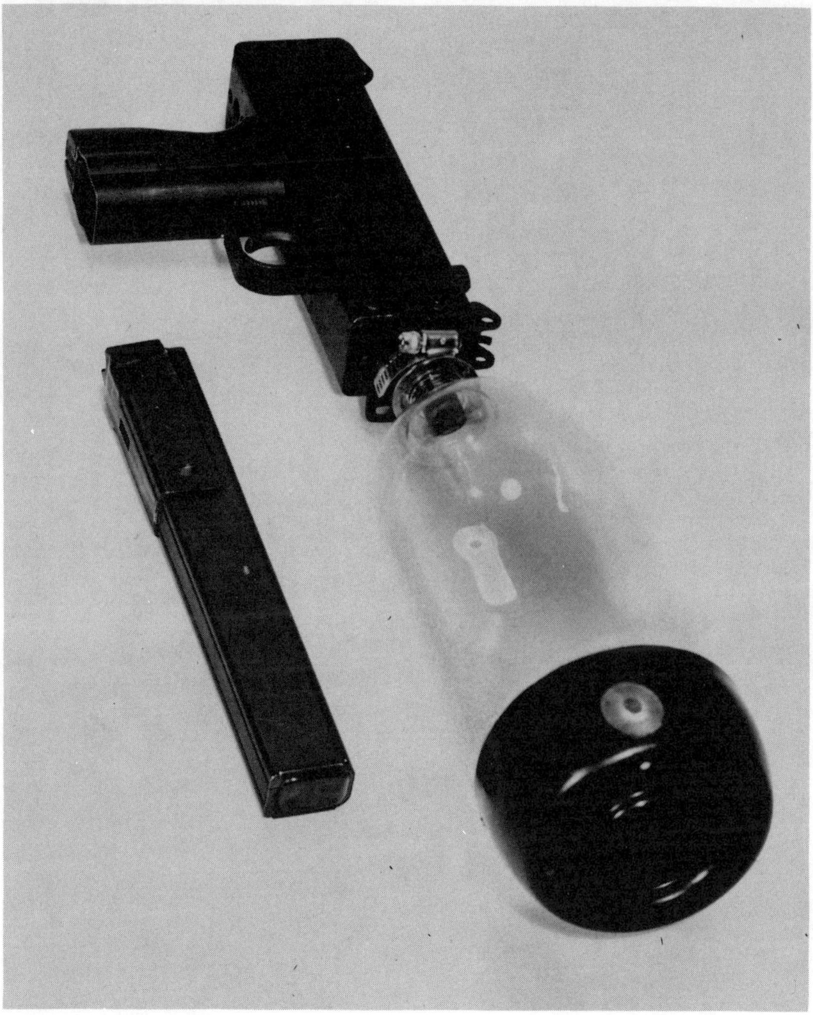

Complete assembly.

Unit can be made more efficient by filling bottle with rags or similar materials.

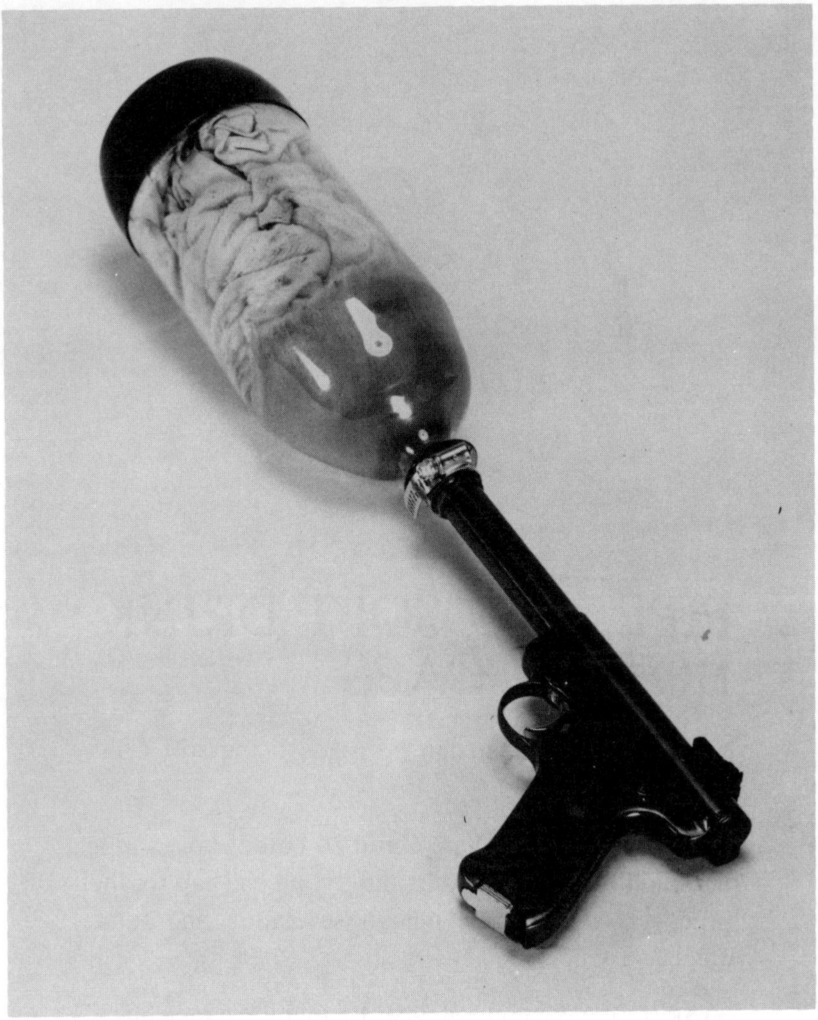

Suppressor attached to Ruger Mk 1 bull barrel .22 pistol.

PIPE AND SOFT DRINK BOTTLE CAPS

A simple silencer can be made using soft drink bottle caps.

Materials required are about 20 bottle caps, a pipe in which the bottle caps fit snugly, an end cap for the pipe, a 1 inch or 1½ inch hose clamp, and some electrical tape.

Start to make your silencer by making a tape bushing behind the front sight. It should be just slightly less in diameter than the inside diameter of the pipe.

Next, make a slot in the pipe for the front sight. Use

a jig saw or hack saw to cut the slot and files to smooth any rough edges. The slot must be long enough so that there is from ½ inch to ¾ inch of pipe extending past the back of the front sight.

Install the pipe to the gun with the hose clamp in the back of the front sight. (Another acceptable method for pipe installation is to slot one end of the pipe and file a notch to one side. This can then be slipped over the front end of the barrel and held in place by twisting the pipe so that it catches behind the front sight).

Proceed by punching and drilling the bottle caps using a drill with a larger diameter than the caliber to be used. Drill end cap making its hole larger than that of the bottle caps.

NOTE: When constructing this disposable silencer, it is best to leave the hole through the center plenty large to prevent the bullet from striking the silencer, if it should become tipped to one side.

Begin filling the pipe with the bottle caps, stacking them all in one direction with the drilled holes pointing towards the end cap. Make sure the last bottle cap has the drilled hole towards the end cap and is half way out of the pipe, that way the end cap will push against it when installed, holding the caps firmly inside the pipe.

Finally, place a cleaning rod or any other suitable rod through the silencer and into the barrel, this is to ensure alignment of the caps with the barrel and end cap. Install end cap. Before firing, remove the cleaning rod.

If you should notice some gas escaping through the slot, use some tape to seal any leaks.

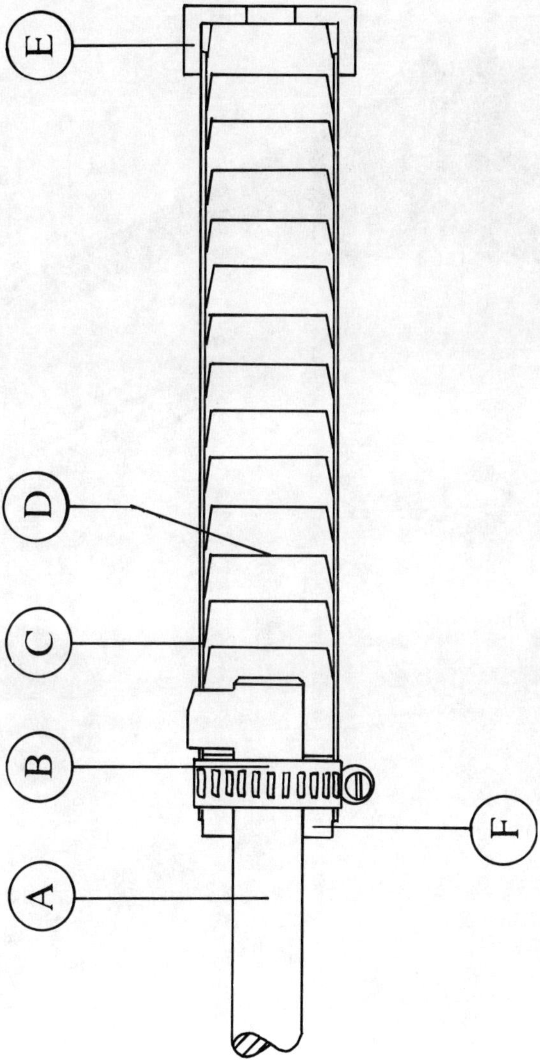

A) BARREL B) HOSE CLAMP C) PIPE D) SOFT DRINK BOTTLE CAPS E) END CAP. F) TAPE BUSHING

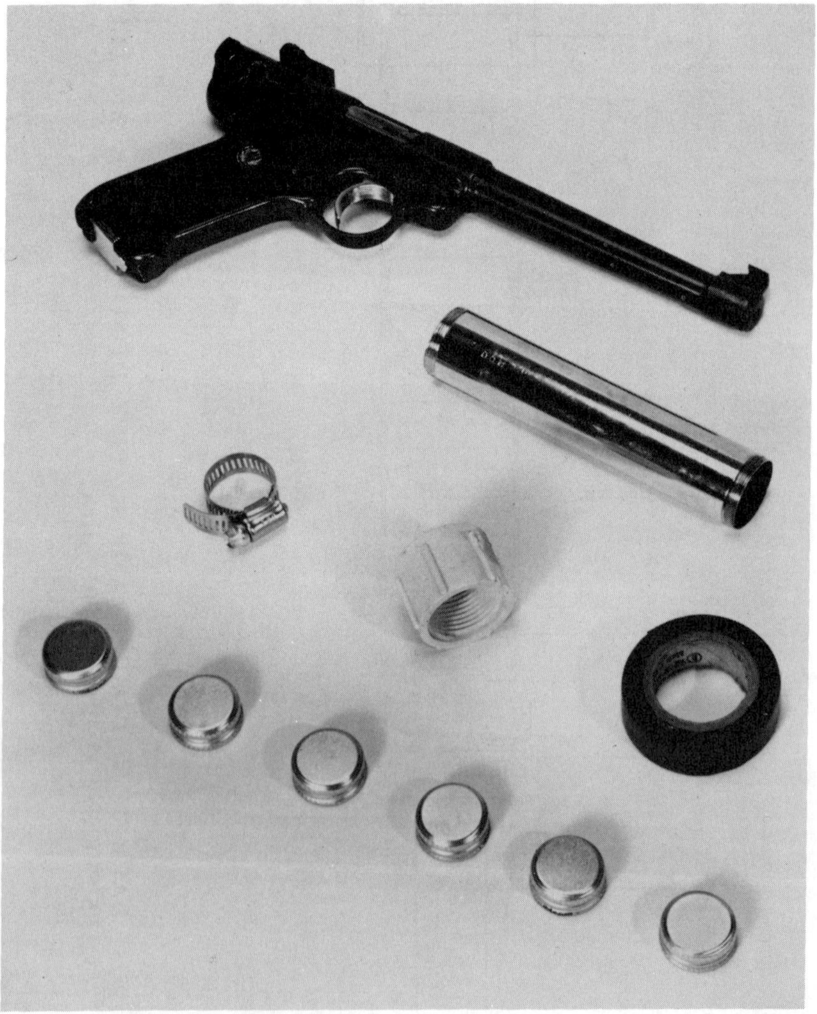

Materials needed: A) PVC or metal pipe B) One inch hose clamp C) PVC end cap D) Electrical tape E) Pop bottle caps.

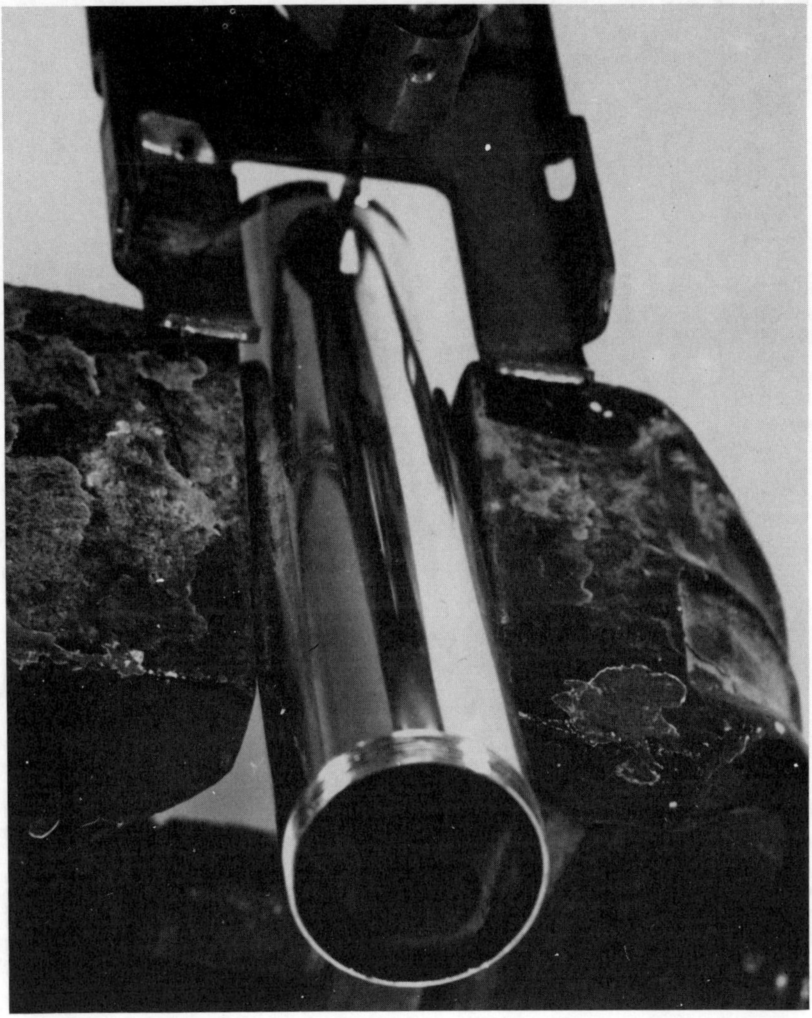

Slotting pipe for front sight.

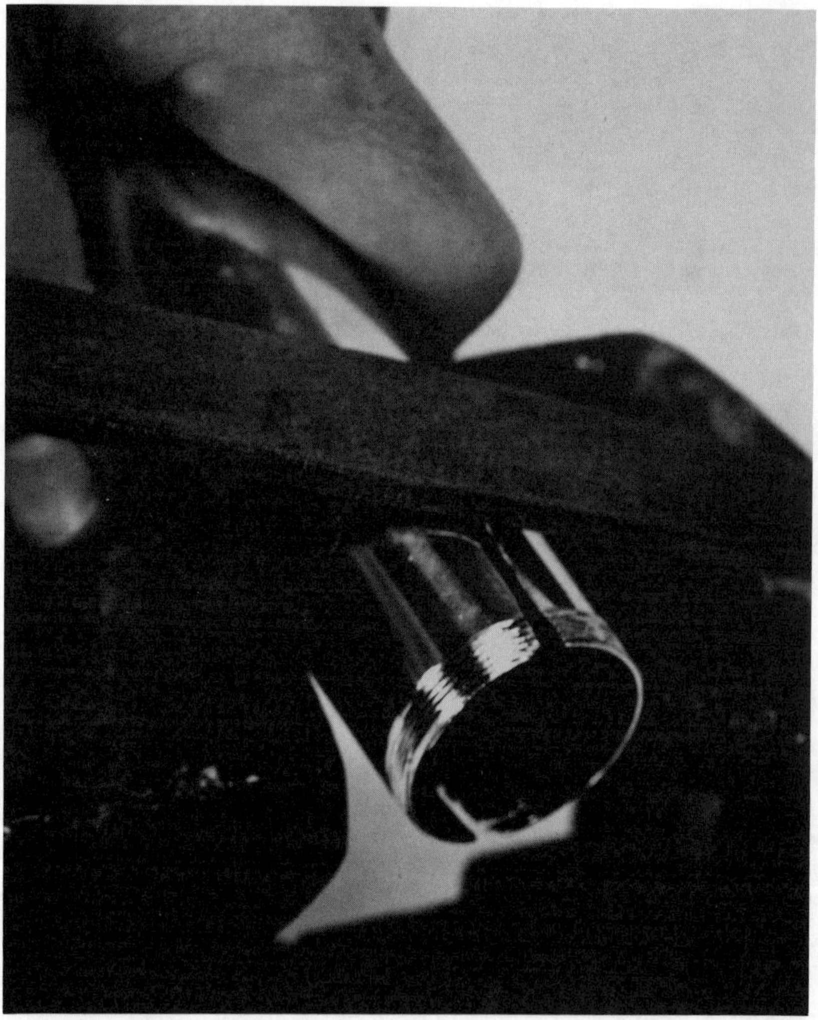

Make sure all rough edges are deburred and smooth.

Finish slot to same width as front sight.

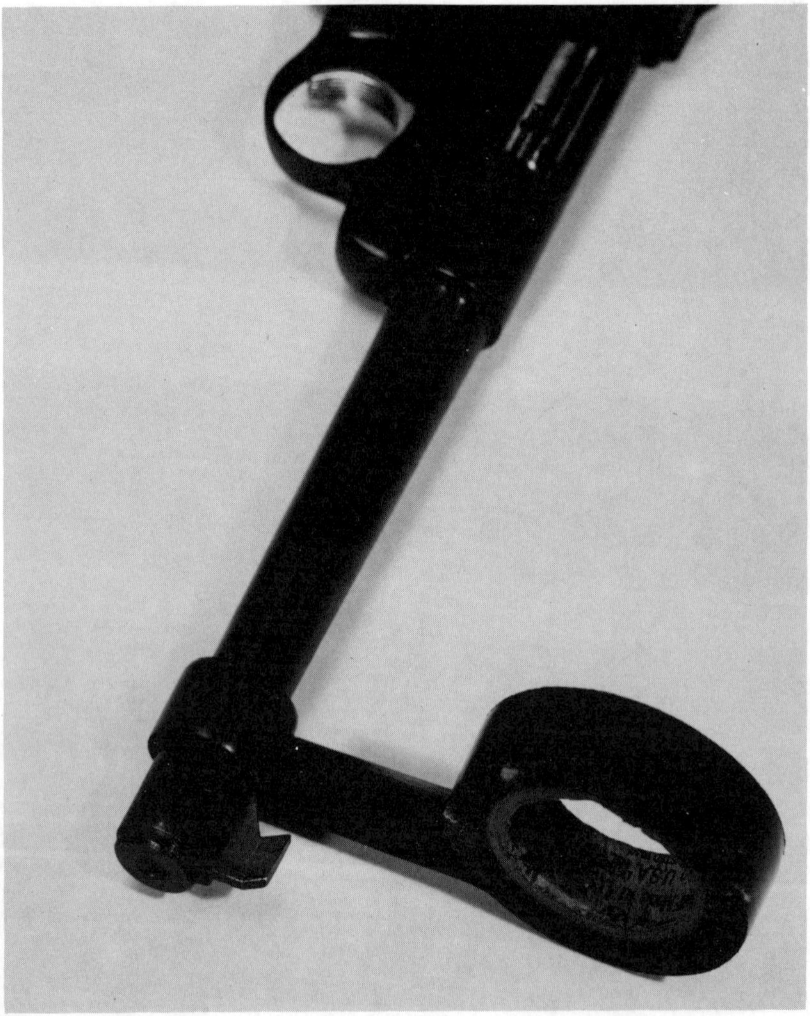

Bring up barrel to a suitable diameter by using electrical tape.

Finished tape bushing.

Use a center punch to mark bottle caps for drilling.

Drill caps using a drill with a larger diameter than the caliber of the bullet to be used.

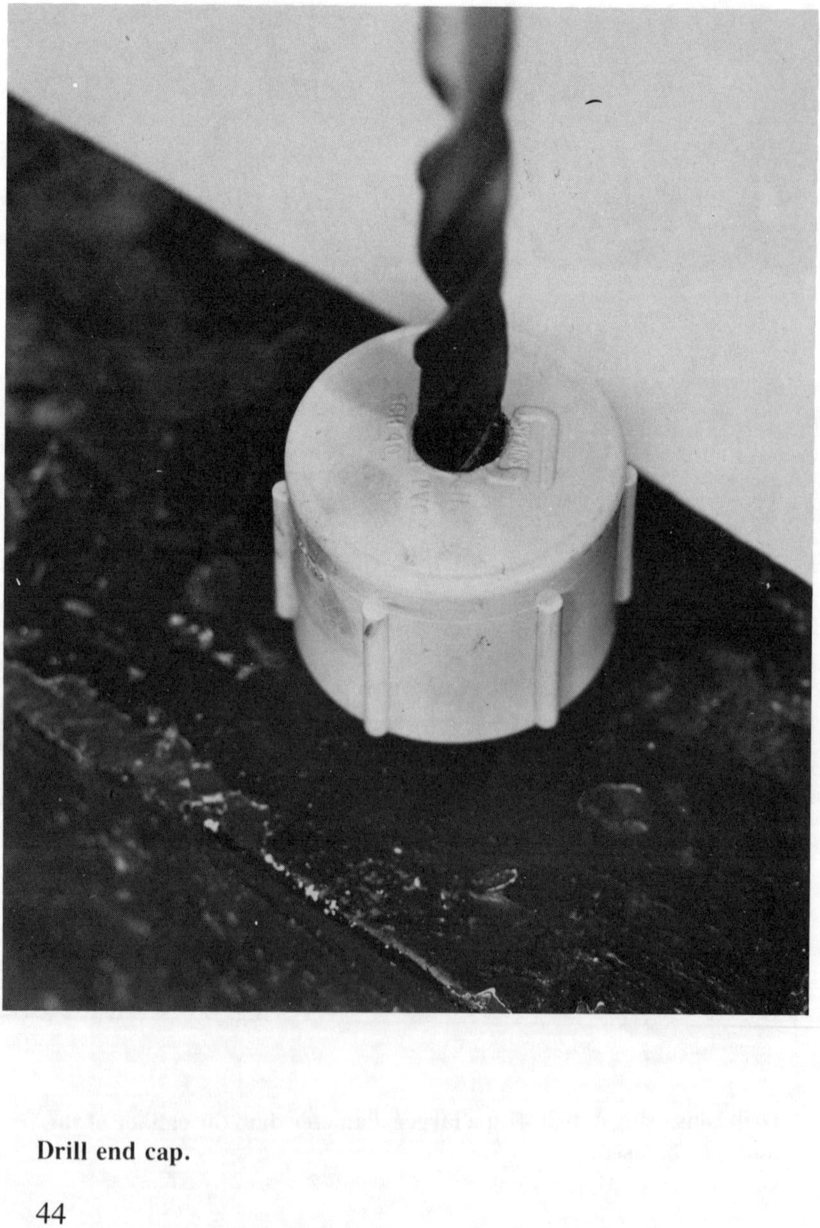

Drill end cap.

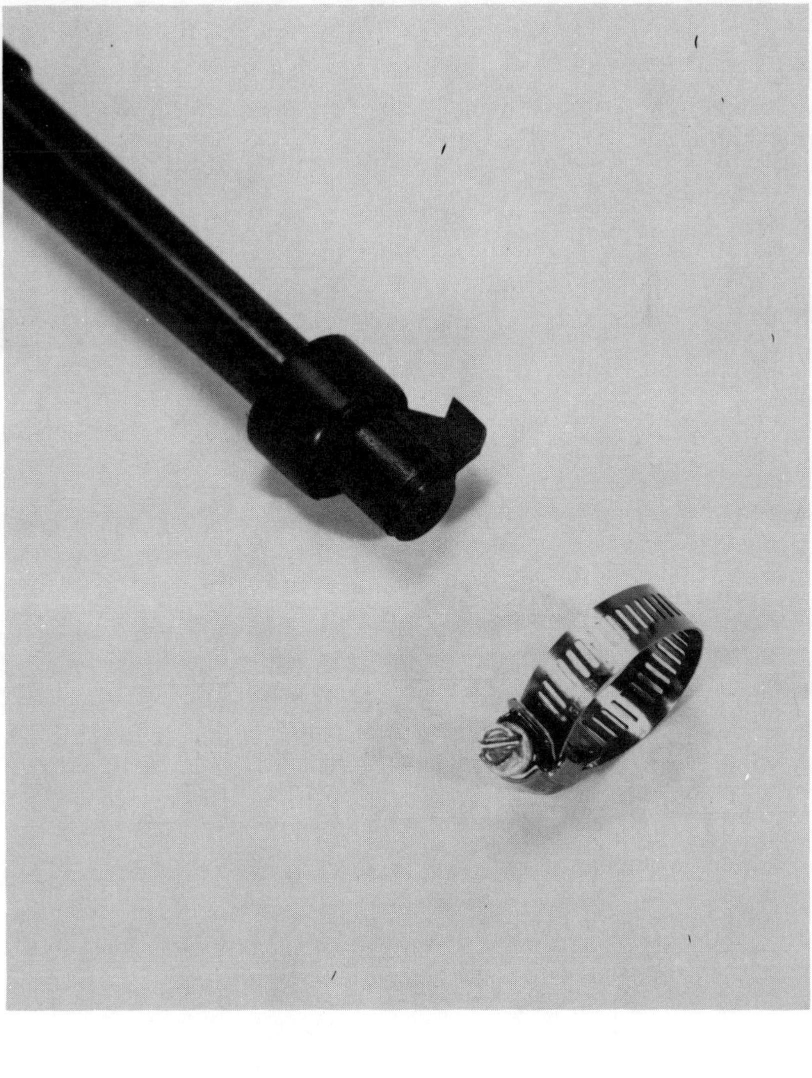

Install hose clamp behind front sight.

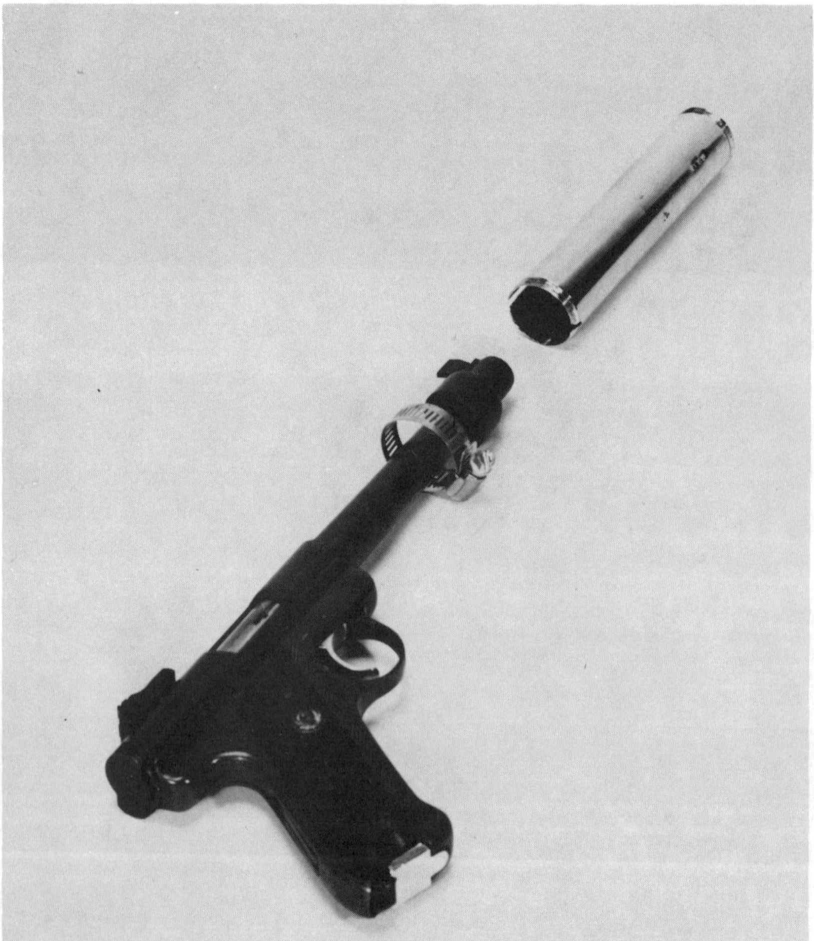

Slip tube over front end of barrel.

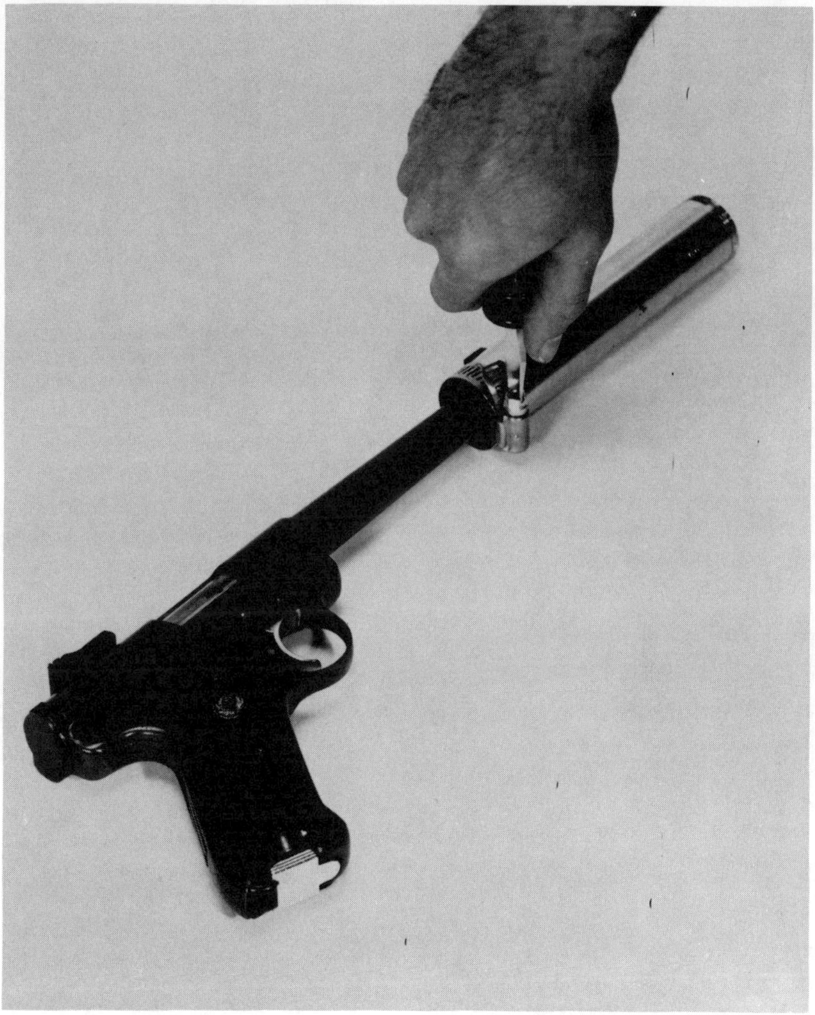

Tighten hose clamp using screwdriver.

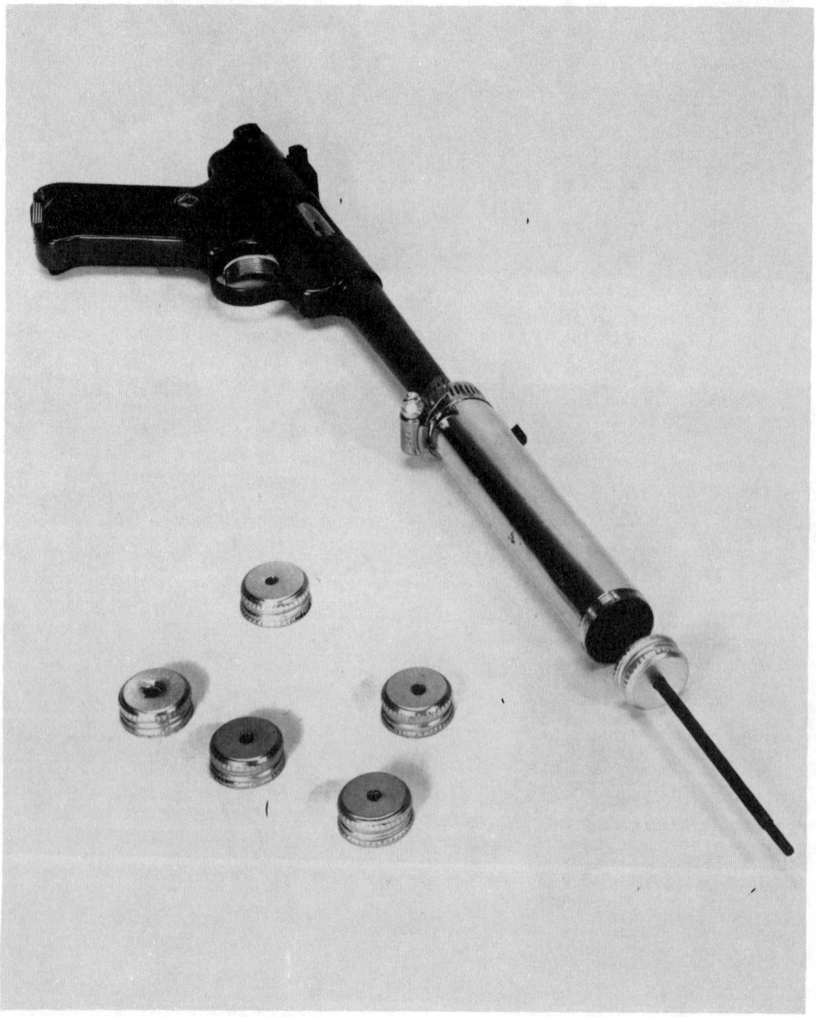

Install bottle caps. Slip cleaning rod into barrel and use it as a guide to align caps.

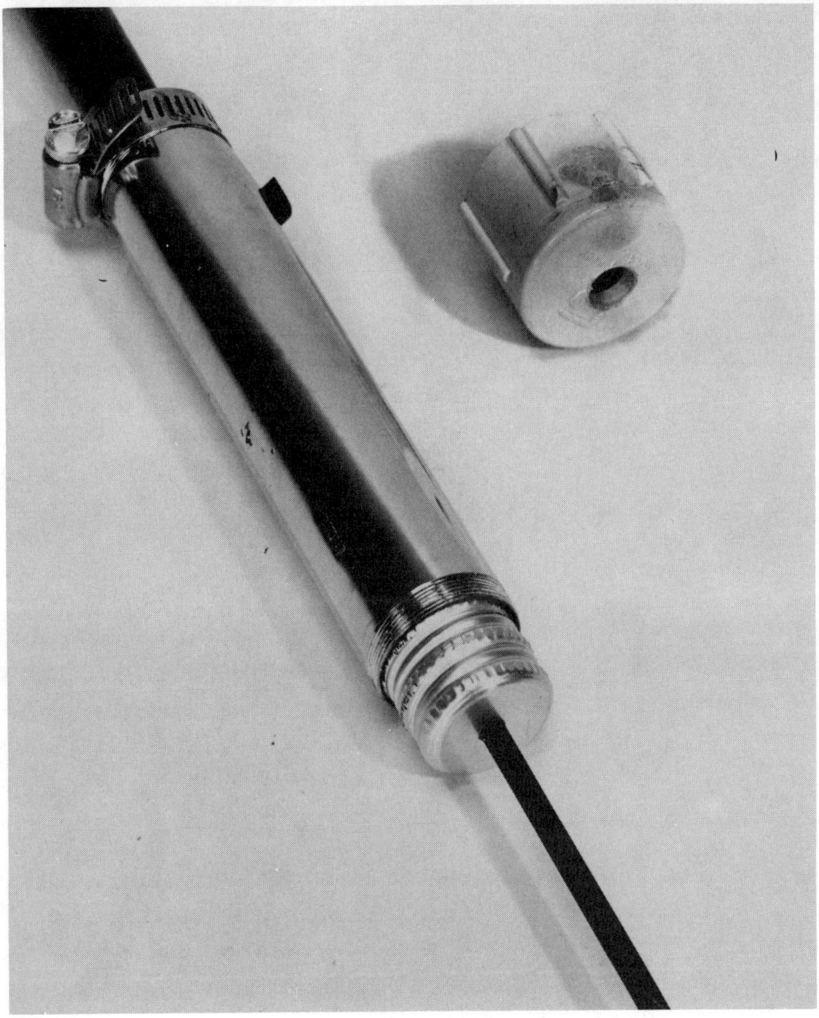

The last cap should be half way out of the pipe, so the end cap will push it tight when installed.

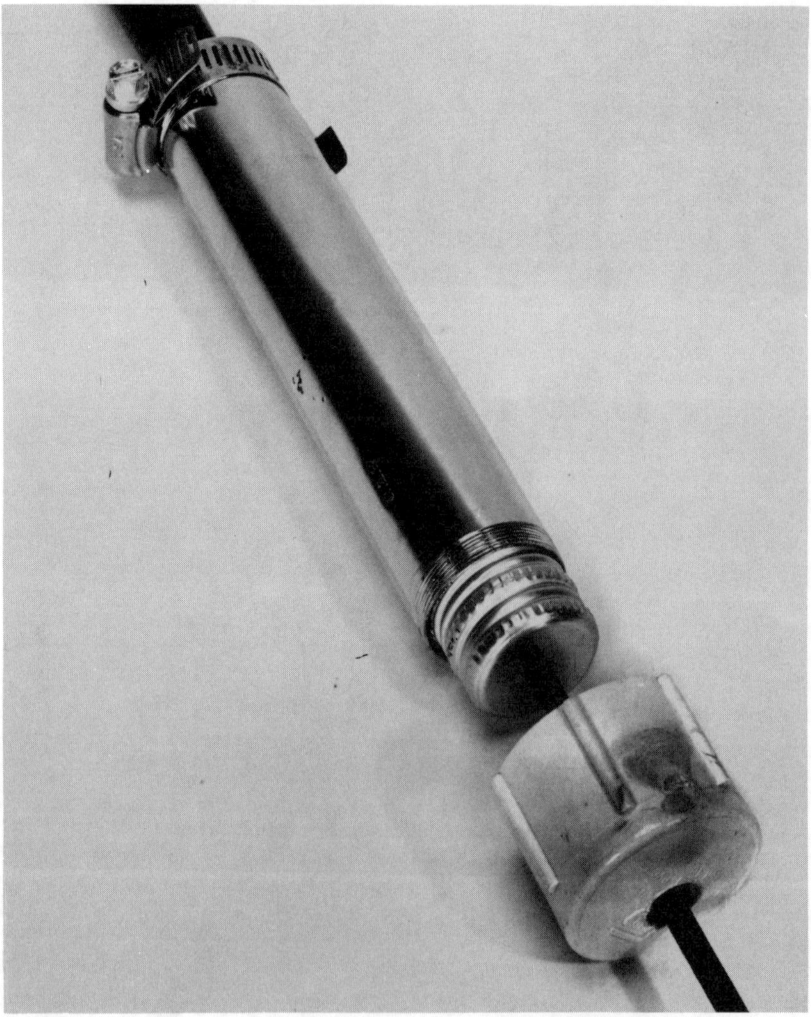

Slip and tighten endcap against bottle caps, making sure it is a tight fit.

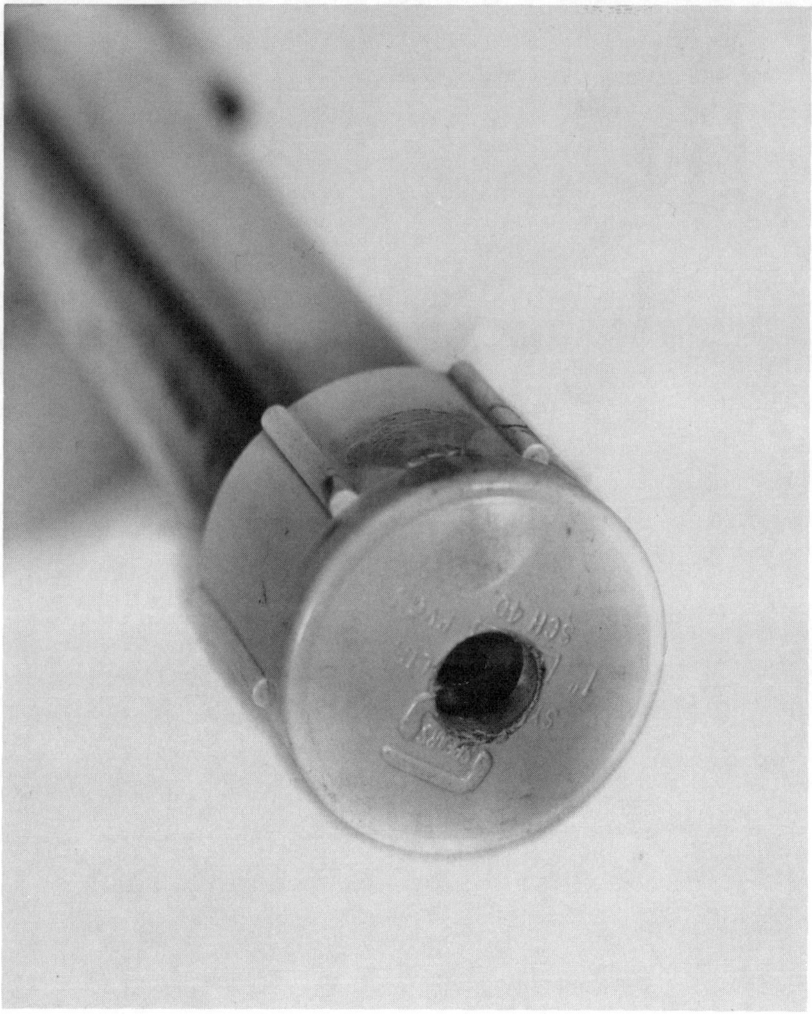

Check to see if holes are aligned inside suppressor.

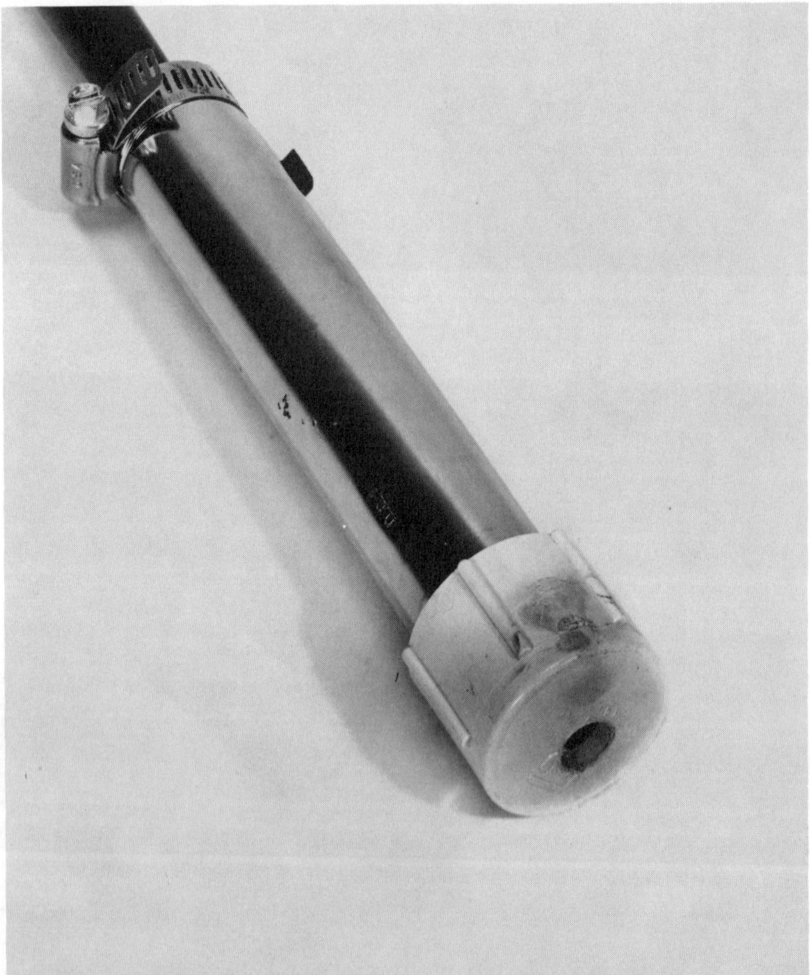

Finished unit ready to be fired.

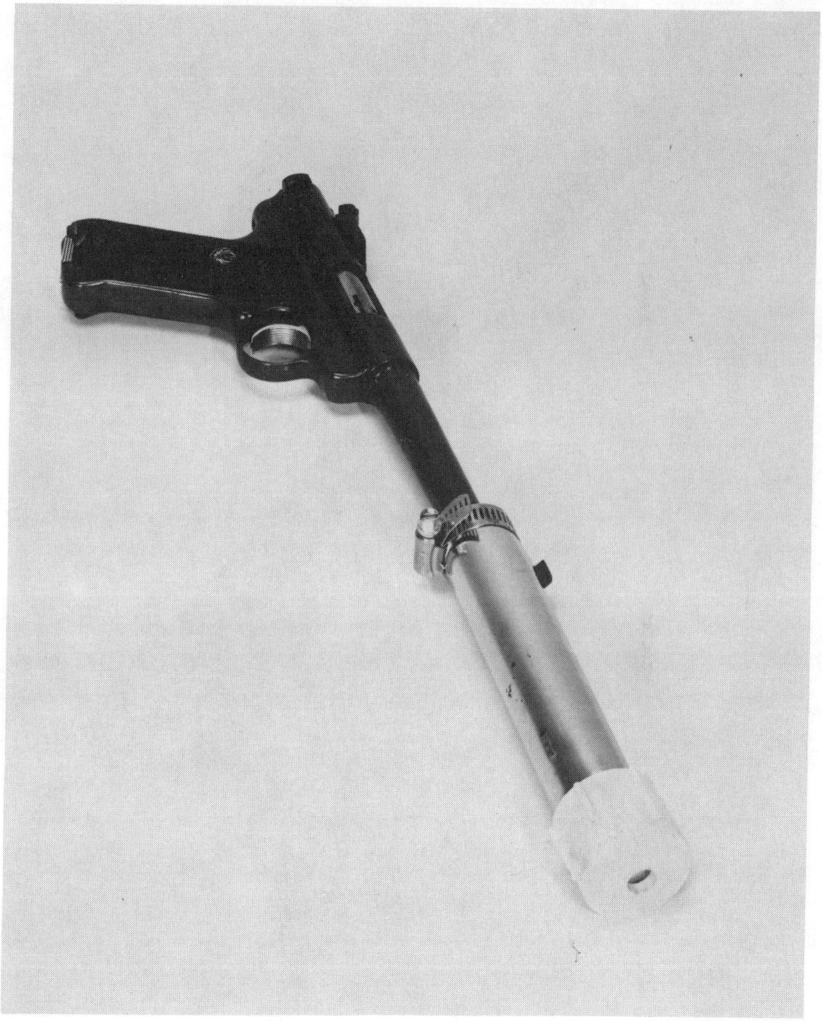

Completed suppressor installed on weapon.

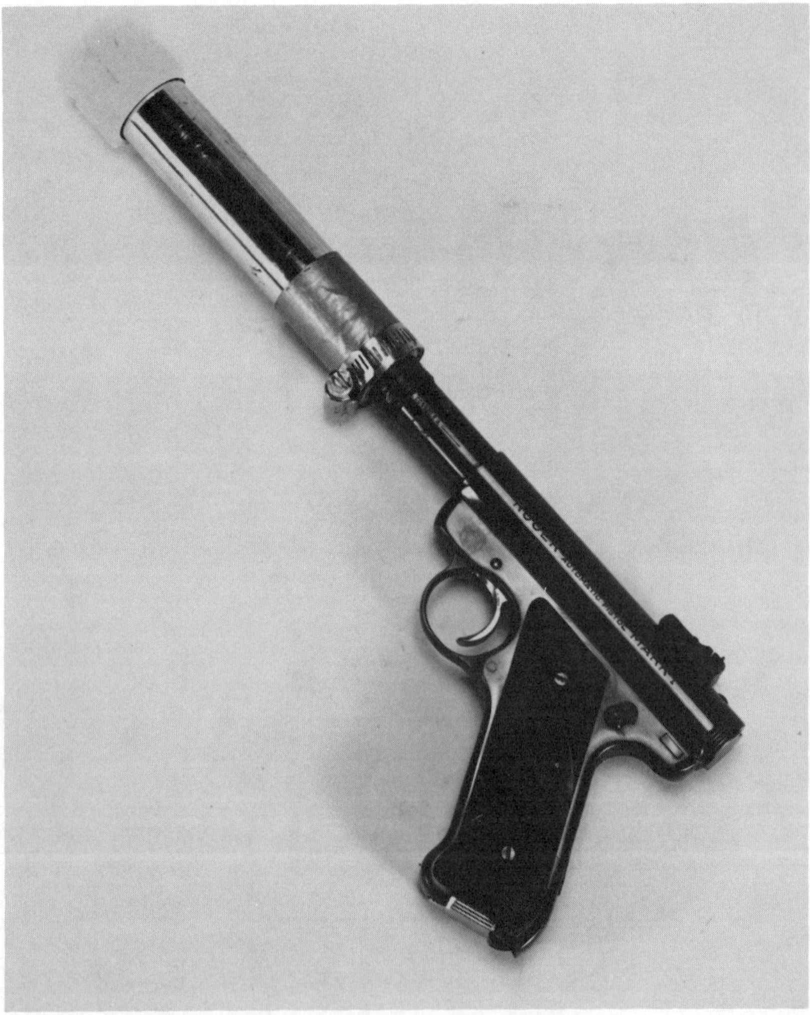

PIPE SILENCER

This one looks more complicated than it really is.

The materials needed for construction are: A piece of PVC or any other type of pipe, about six or seven inches long, with the same inside diameter as the outside diameter on your weapon's barrel. (A diameter slightly larger will work if tape is used to bring up the barrel to a suitable diameter.) A one inch hose clamp, a piece of soft rubber, copper scouring pads (available at most supermarkets as "CHORE BOY PURE COPPER SCOURING PADS") and some wide electrical tape.

Begin construction by drilling the pipe with four rows of evenly spaced ¼ inch diameter holes, 90 degrees apart. The number of holes will be determined by the length of the pipe. (Deburr all holes, both inside and outside after drilling.)

Next, cut a slot for the front sight, using a jig saw or hack saw for rough cutting and files to smooth the edges. The slot should be long enough so that the pipe extends about ½ inch to one inch past the back of the front sight. Try to make it as tight a fit as possible.

Slide the pipe onto the barrel, with the front sight in the slot. Position the hose clamp behind the front sight and close to the end of the pipe. Tighten clamp (use tape to bring up barrel to size, if necessary). Make sure that the pipe is snug and tight against the barrel to prevent it from wobbling.

Cut the piece of soft rubber into a disk shape having a diameter slightly larger than that of the outside diameter of the pipe. With a sharp razor or knife make an "X" at the center of the disk. Put the rubber disk aside for the moment.

Cut out the rivet in the copper scouring pads and unfold them, this will make it easier to wrap them over the pipe.

Wrap the copper scouring pads (several should be used, depending on the length of the pipe), around the pipe as tightly as possible and hold them in place with the palm of your left hand. Place the soft rubber disk with the "X" at the center of the pipe hole, and hold in place with the forefinger and thumb of your left hand. Begin taping around the rubber disk and roll of copper pads with your right hand, making certain the "X" in the rubber disk is still aligned with the hole in the pipe. (A simple way to assure alignment is to pass a cleaning rod through the "X" and into the barrel.) Using wide tape will make the job a lot easier. Extend the tape far enough to the front so it will hold the rubber disk in place.

Once the rubber disk is in place, begin wrapping back, overlapping the tape as you go. Extend the tape back to the front sight covering the whole "silencer" with two layers. Be certain no sections of the copper scouring pads are exposed.

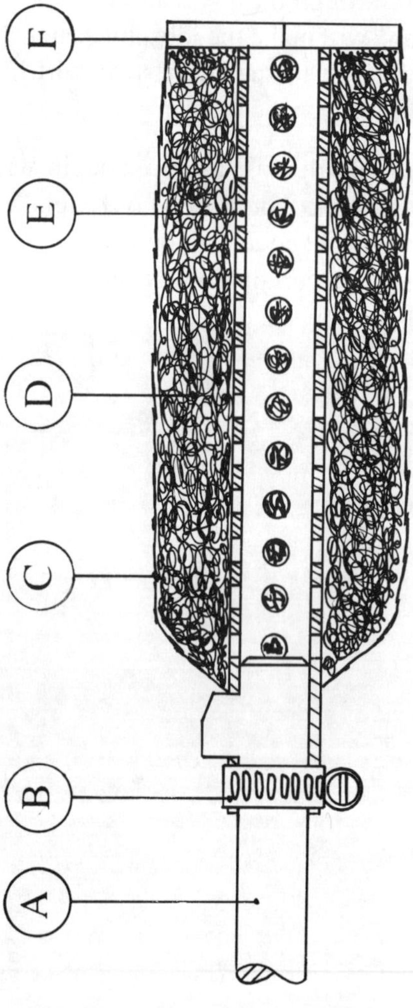

A) BARREL B) HOSE CLAMP C) ELECTRICAL TAPE D) COPPER SCOURING PADS E) PIPE F) SOFT RUBBER DISK

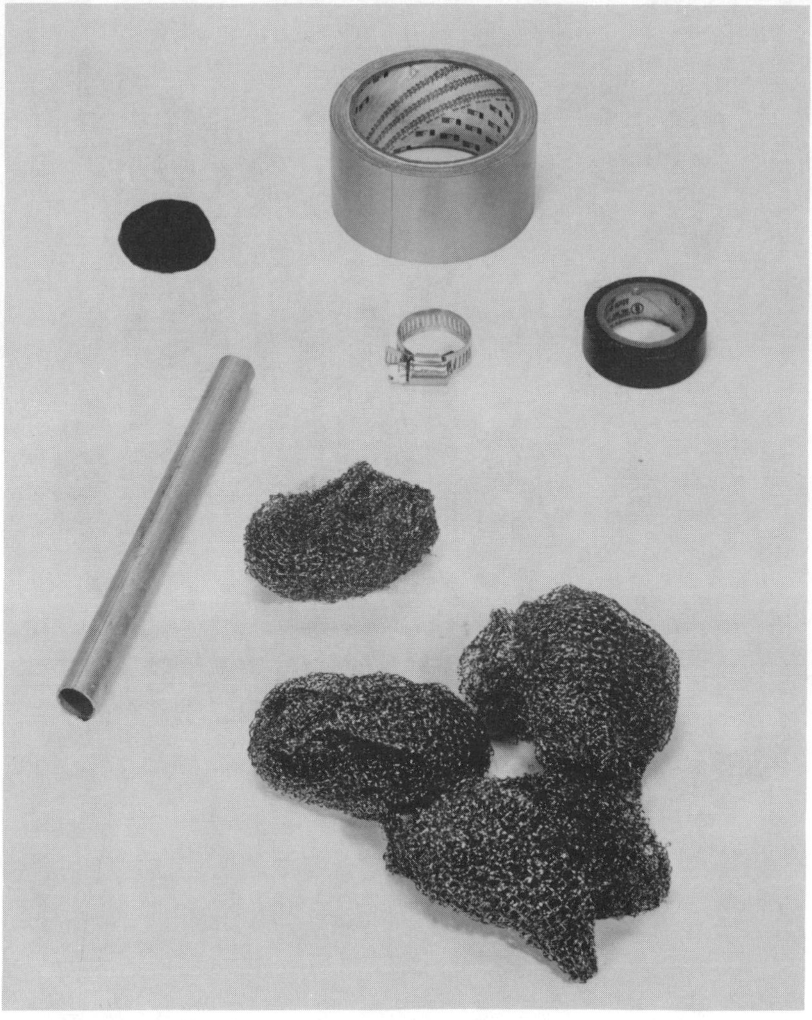

Materials needed: A) piece of rubber B) Cloth adhesive tape C) Electrical tape D) One inch hose clamp E) Pipe F) Copper scouring pads.

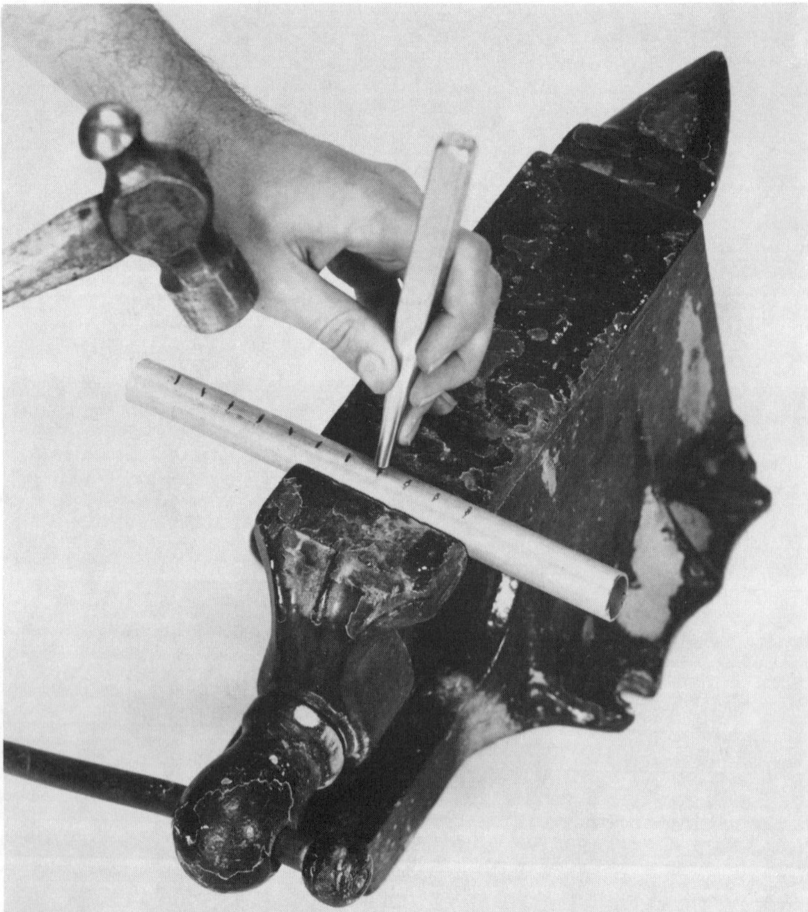

To make drilling easier, mark pipe using a center punch (for dimensions, see text).

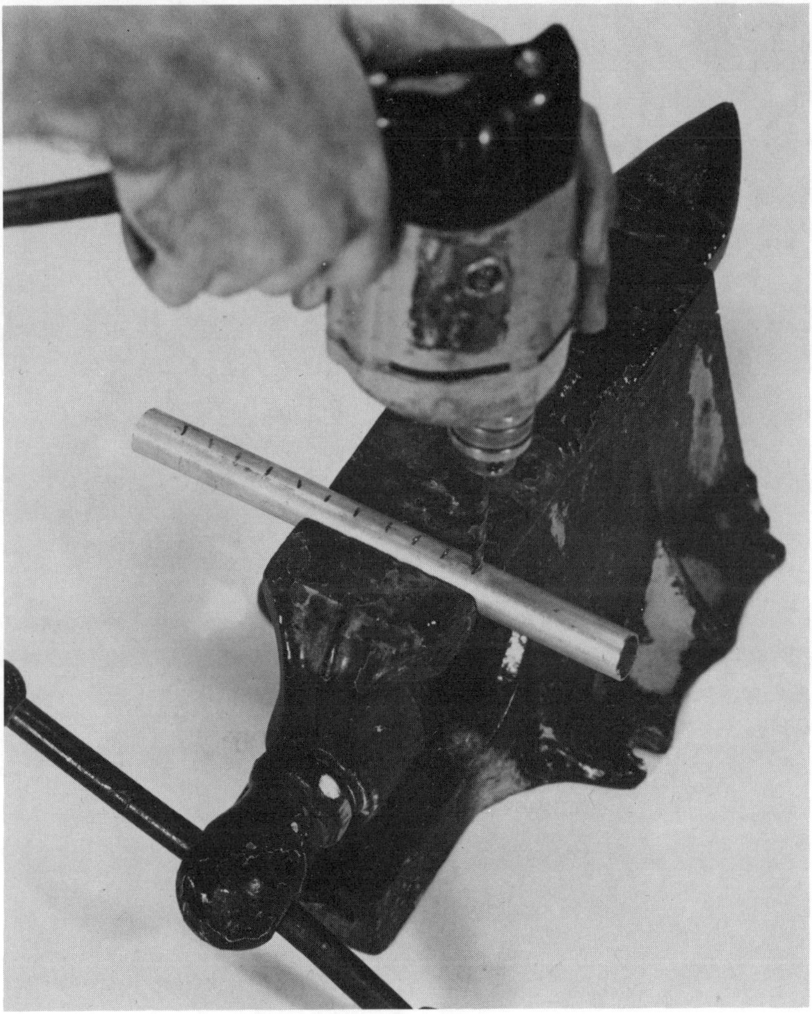

Drill holes using hand drill or drill press.

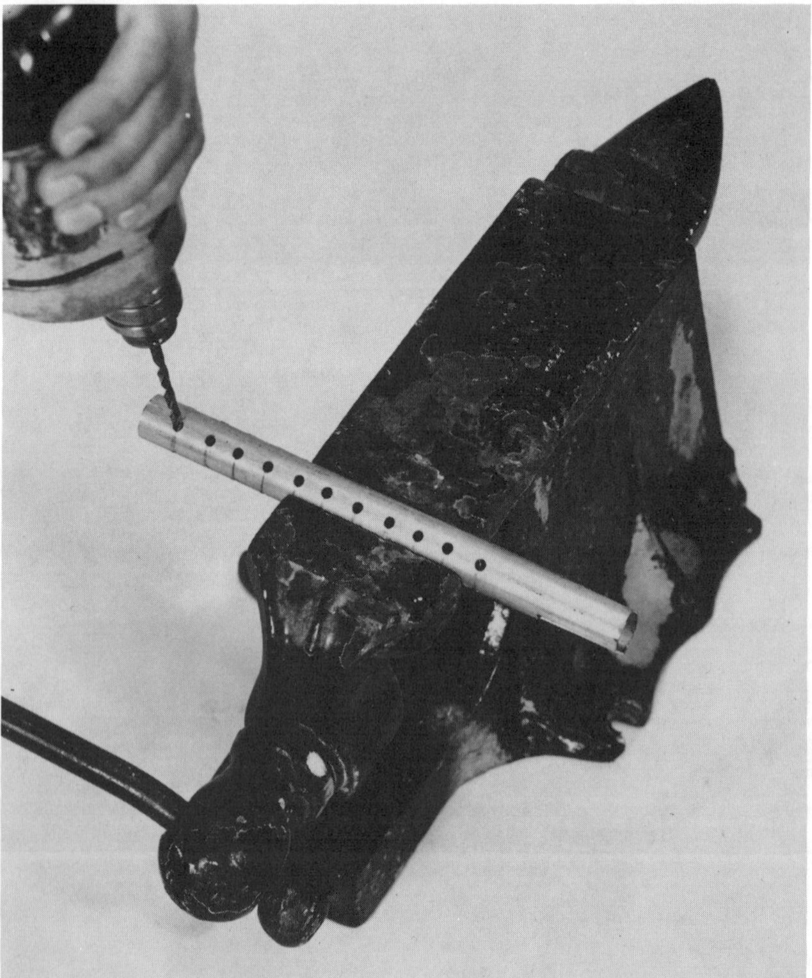

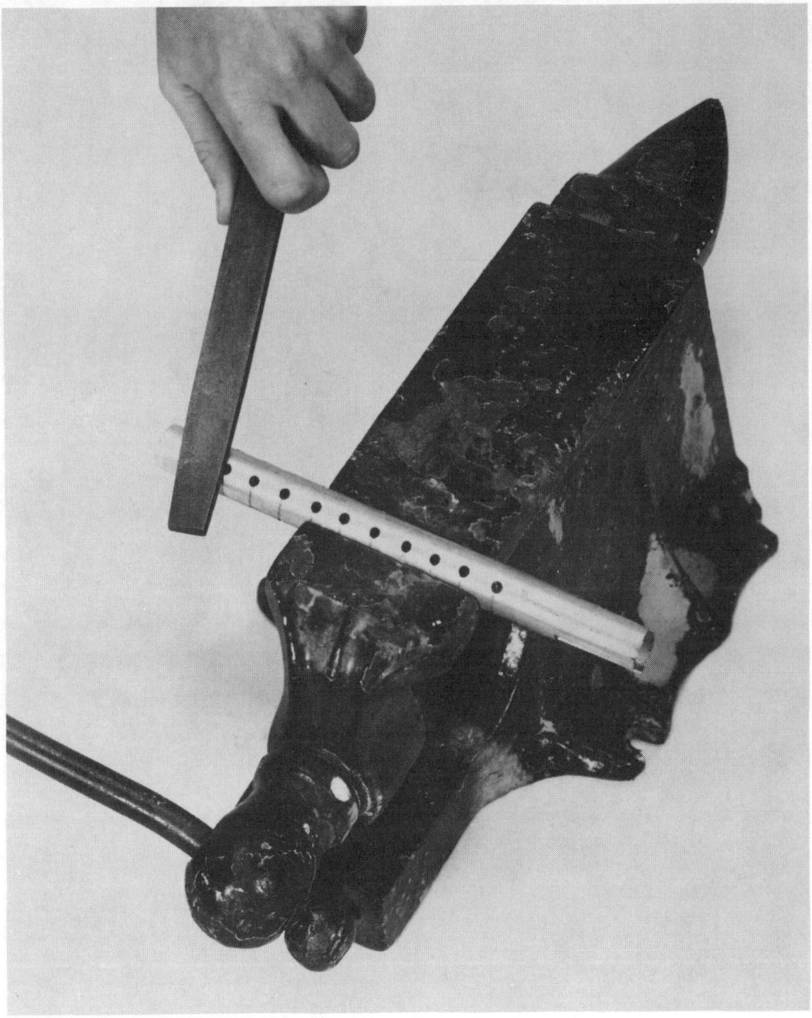

Deburr all holes inside and out.

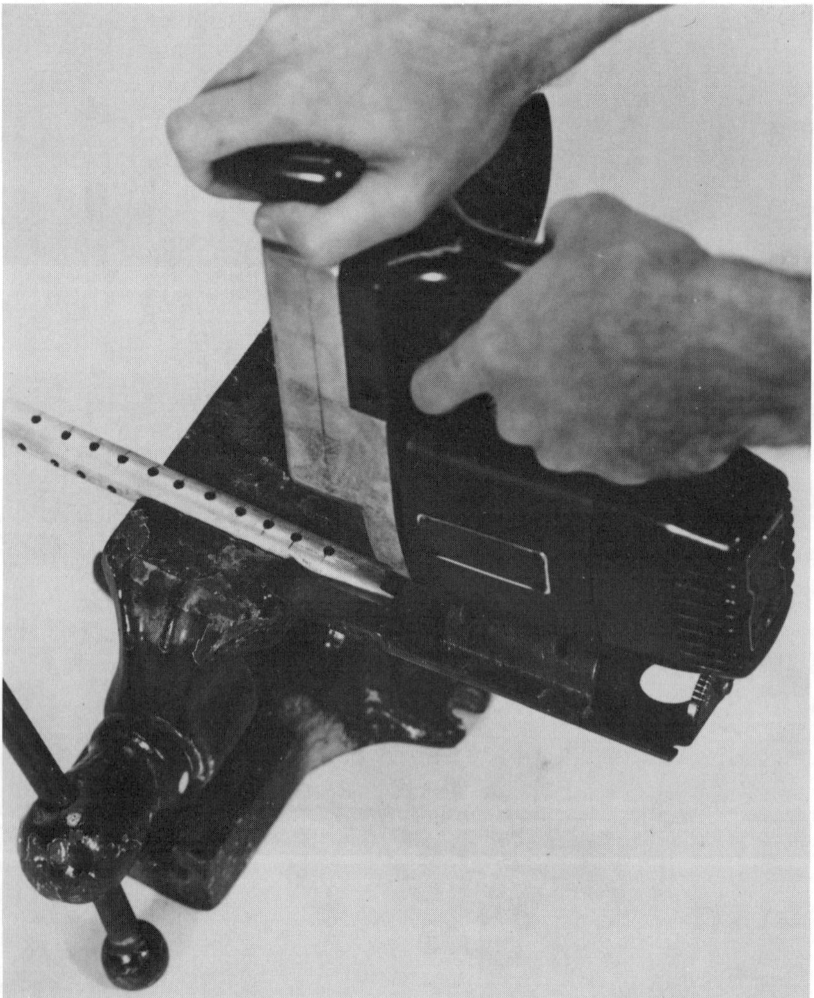

Cut a slot for the front sight. Use hacksaw or jigsaw if available.

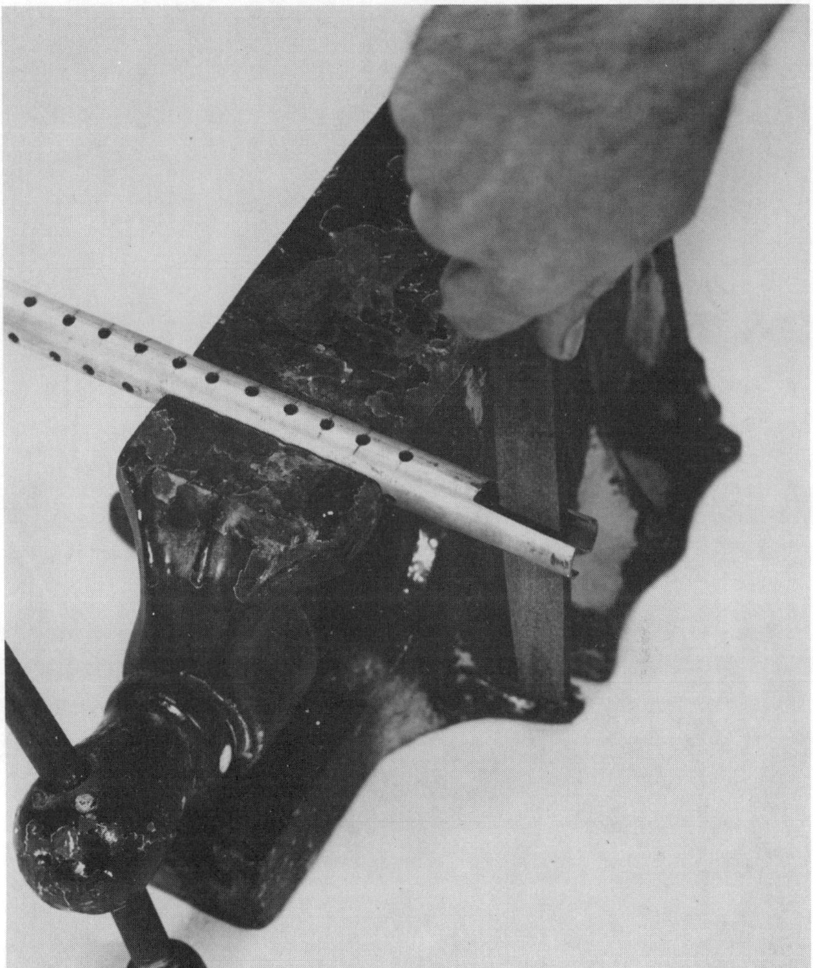

File slot to size to accommodate front sight.

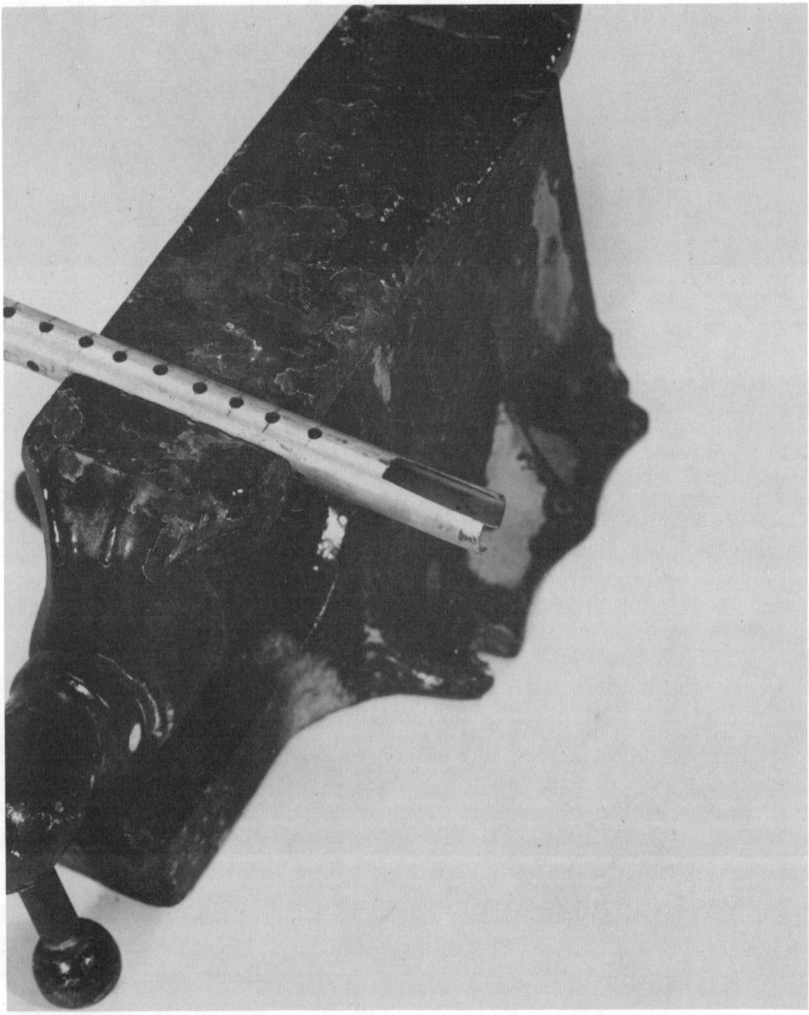

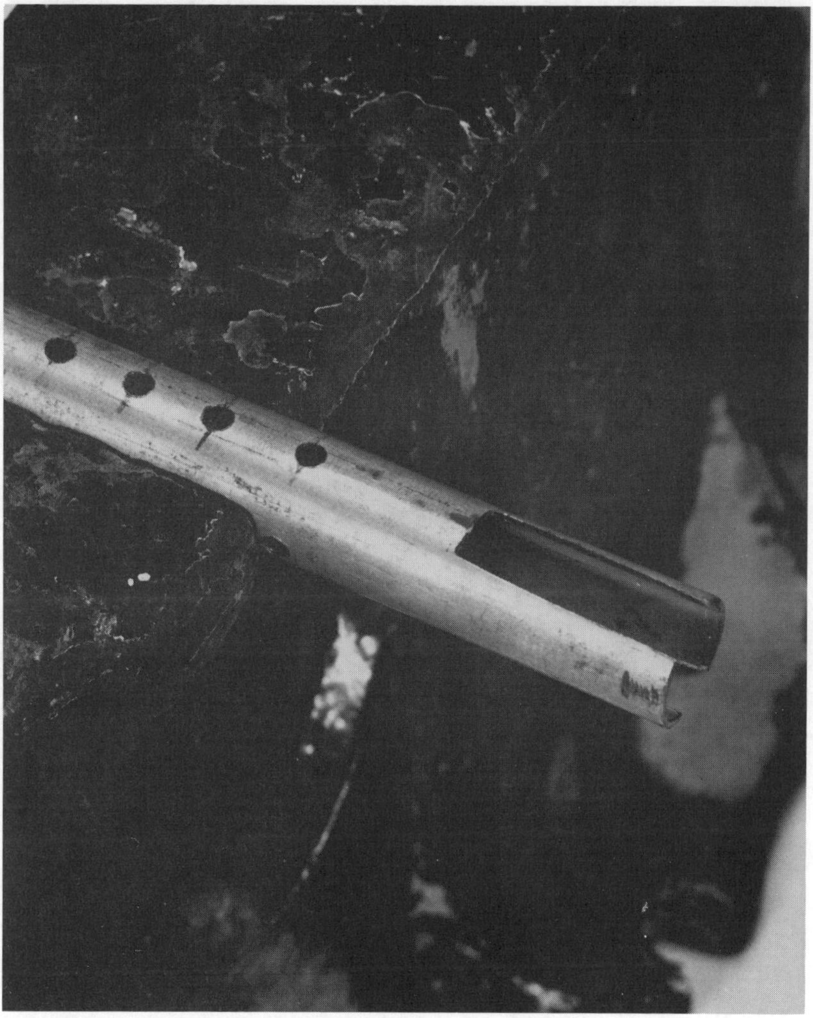

Finished front sight slot.

Bring barrel up to the same size as the inside diameter of the pipe.

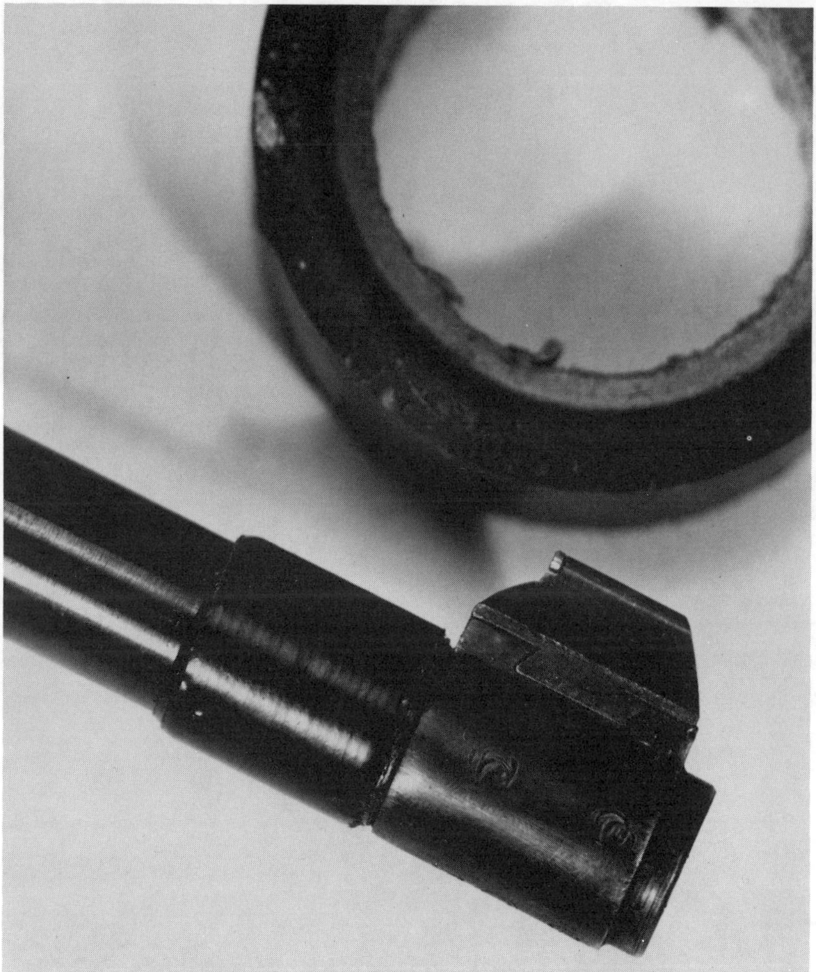

Slip hose clamp behind front sight.

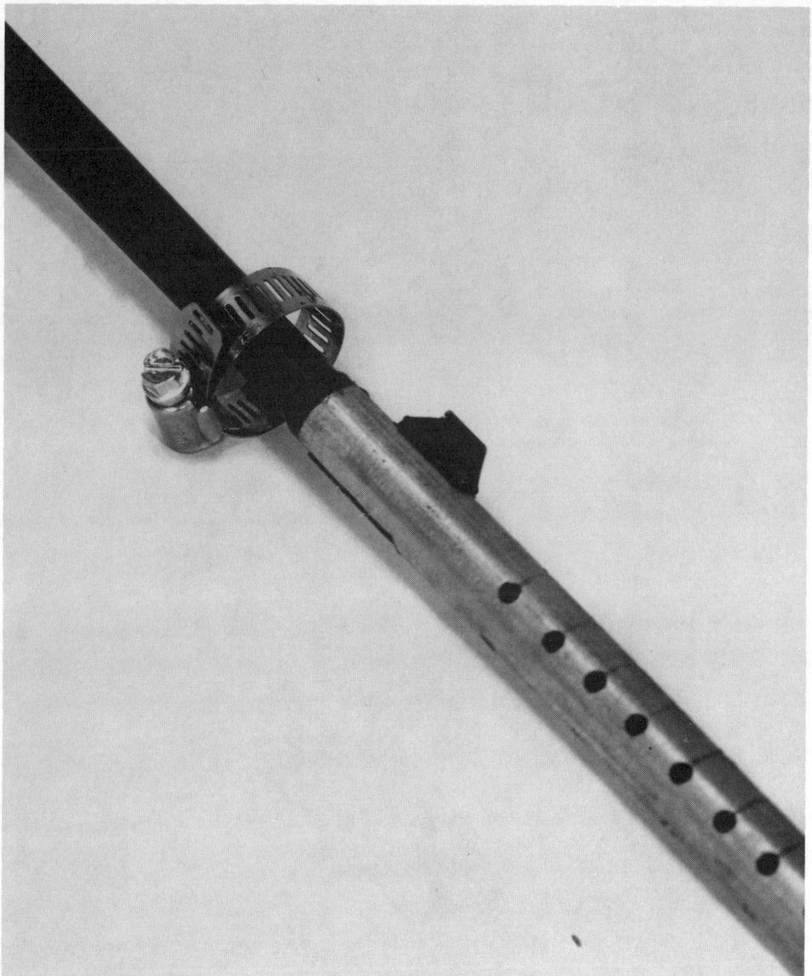

Install pipe on barrel.

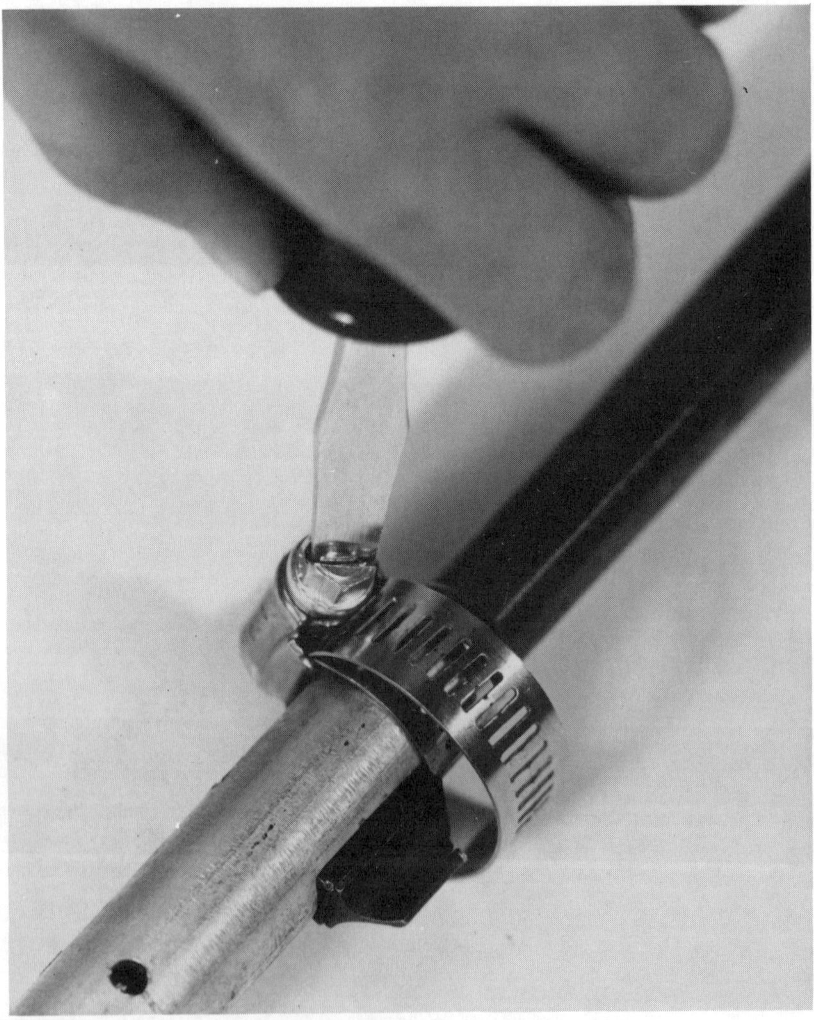

Tighten hose clamp behind front sight to hold pipe in place.

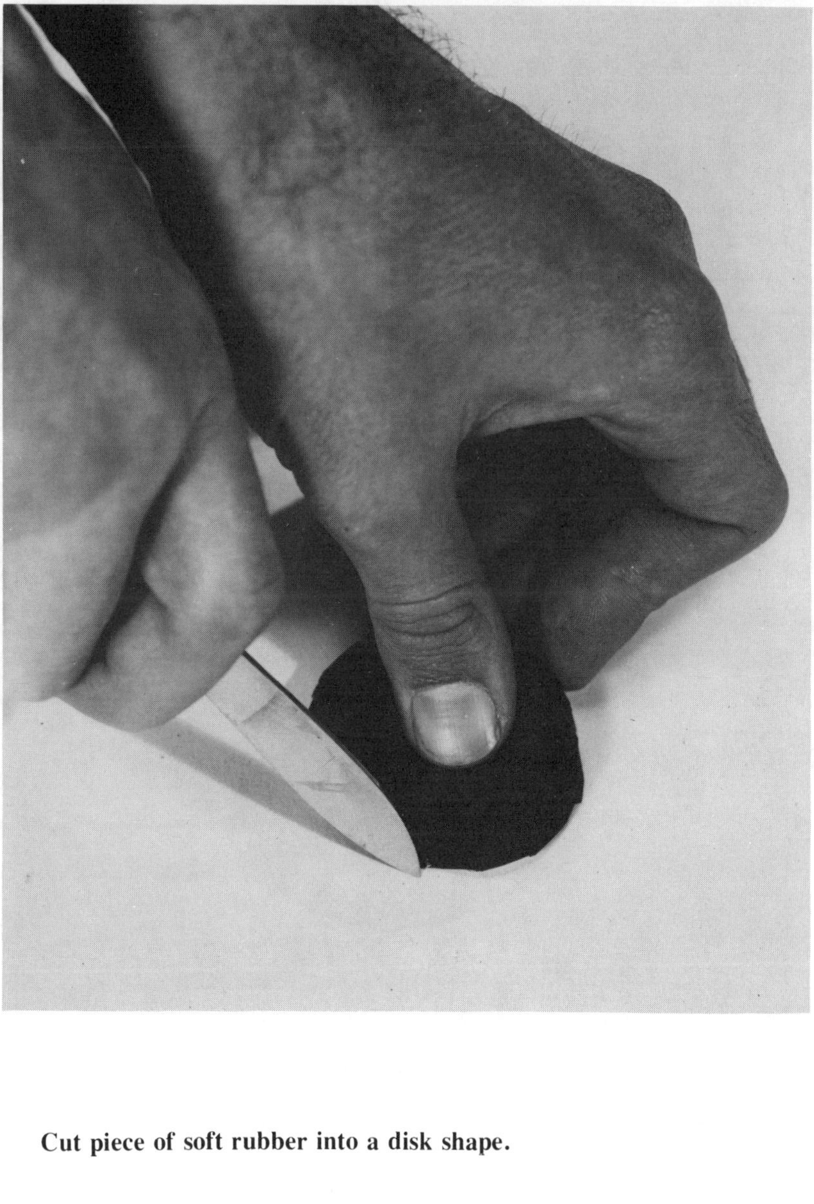

Cut piece of soft rubber into a disk shape.

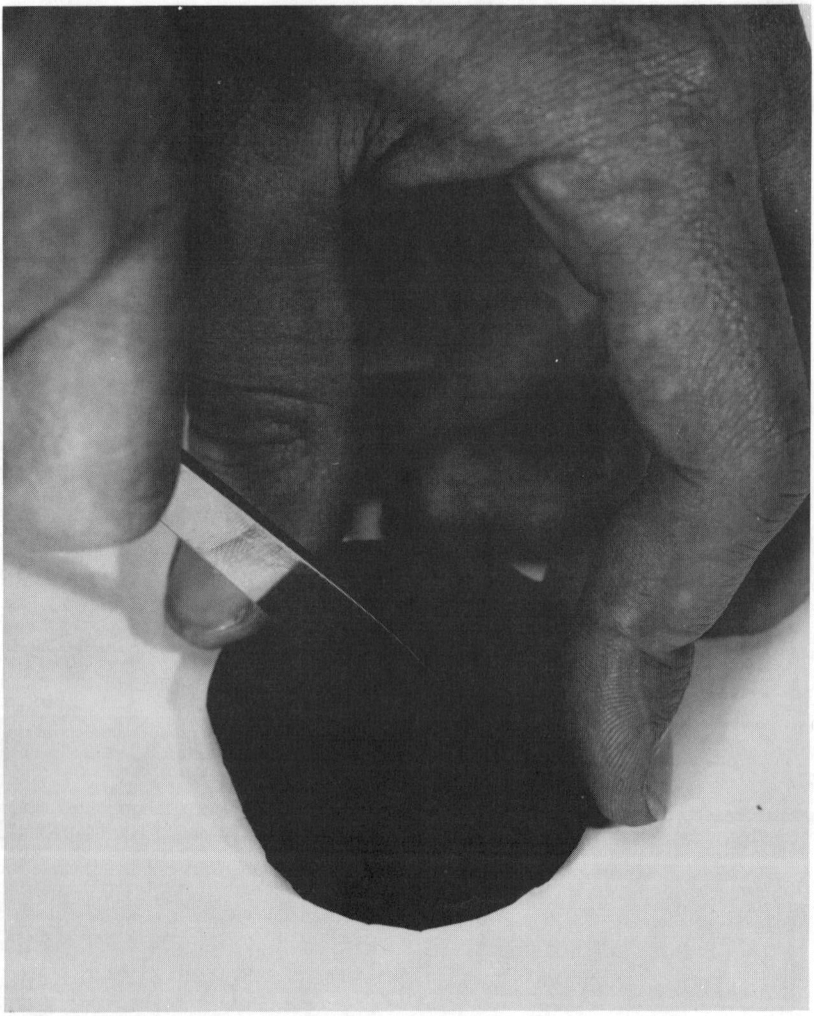

Use sharp razor or knife to cut an "X" at the center of the rubber disk.

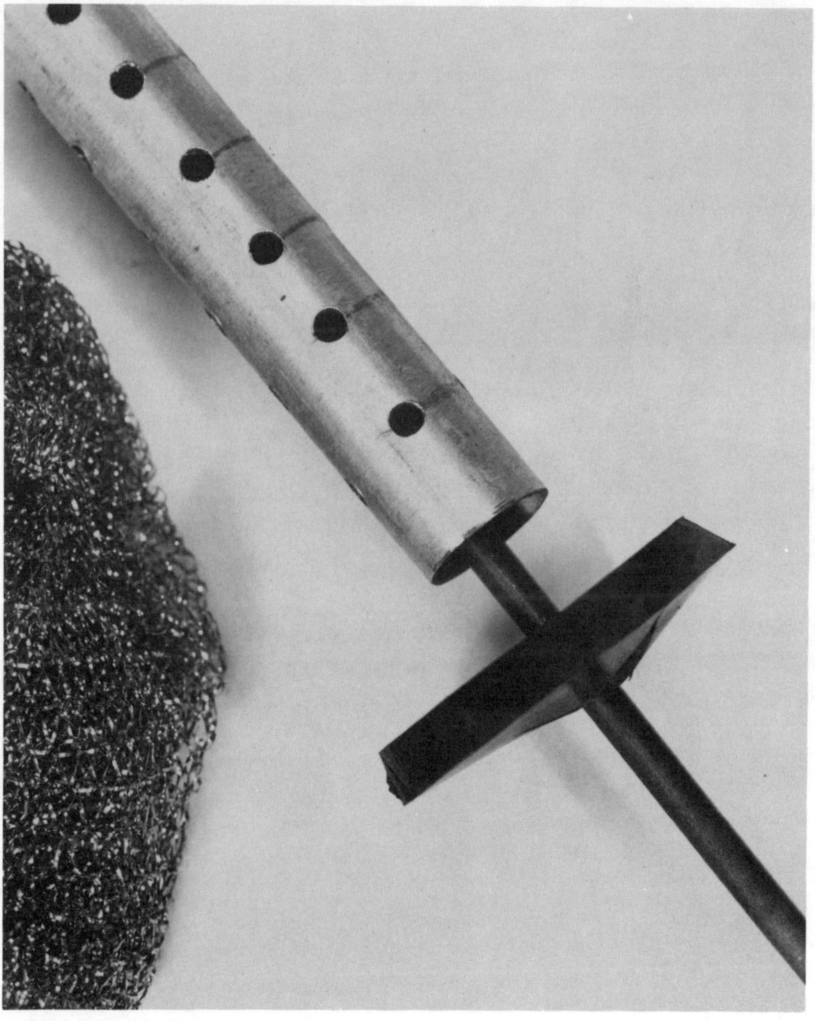

Slip cleaning rod through pipe and into barrel, then slip rubber disk.

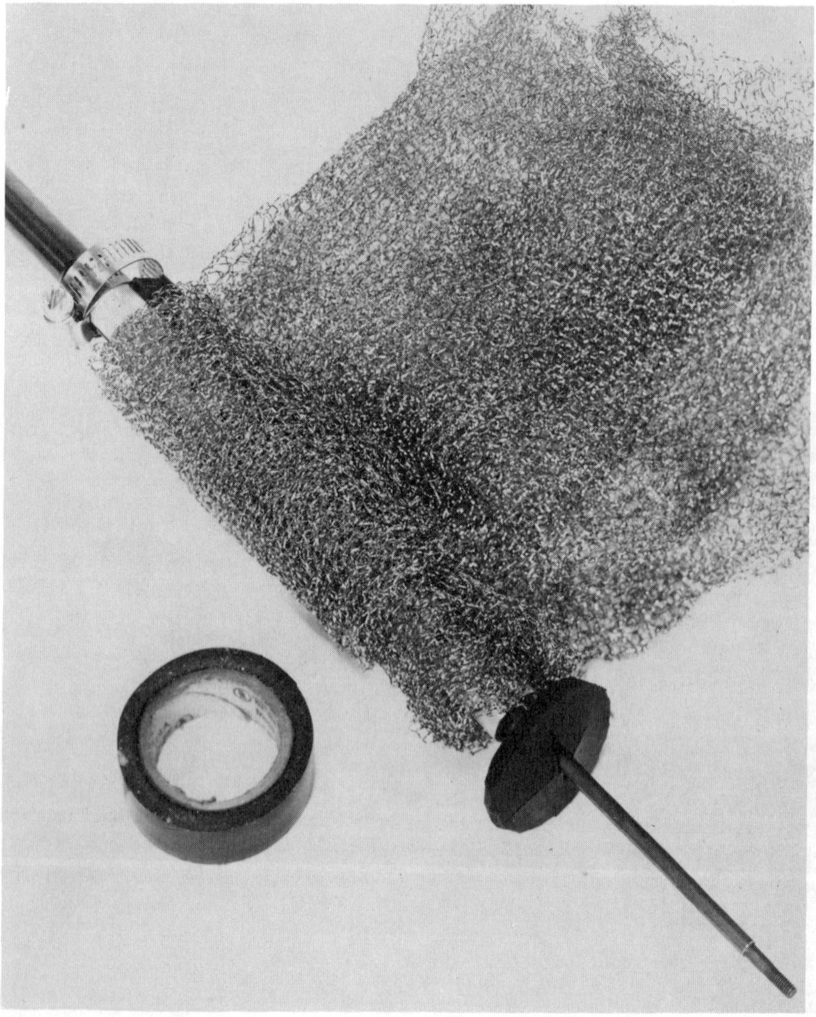

Start wrapping copper scouring pads over pipe.

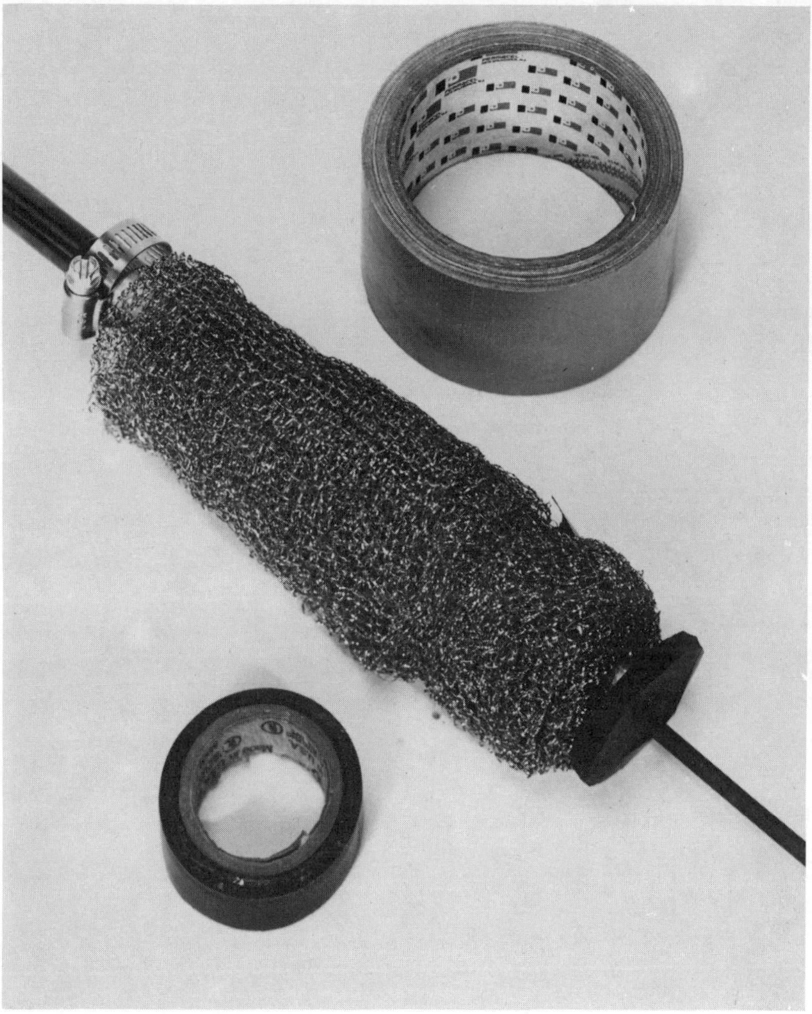

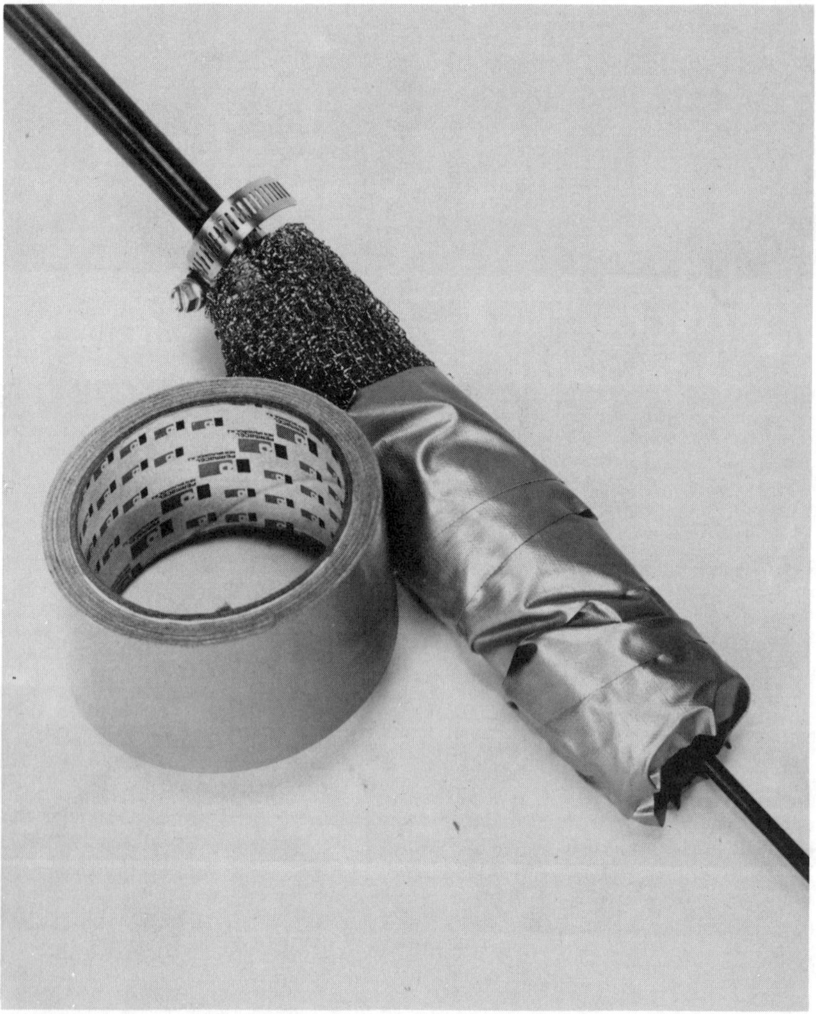

Wrap entire unit with two layers of cloth adhesive tape.

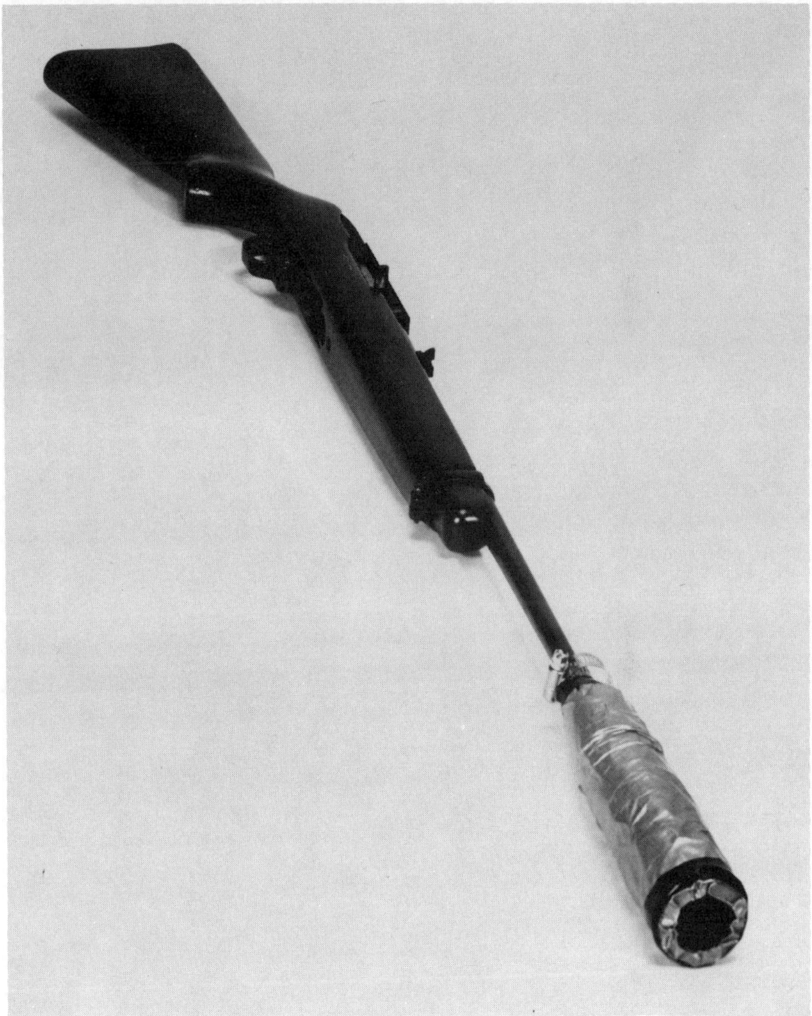

Completed suppressor mounted on Ruger 10/22 .22 caliber rifle.

Notes

Notes

Other Books Available From Desert Publications

No.	Title	Price
081	Crossbows/ From 35 Years With the Weapon	$15.95
206	CIA Field Expedient Preparation of Black Powder	$19.95
221	Evaluation of Improvised Shaped Charges	$19.95
411	Homeopathic First Aid	$14.95
454	Survival Childbirth	$15.95
C-002	How To Open A Swiss Bank Account	$9.95
C-011	Defending Your Retreat	$9.95
C-020	Methods of Long Term Storage	$9.95
C-023	Federal Firearms Laws	$11.95
C-024	Select Fire Uzi Modification Manual	$13.95
C-025	How to Build Silencers	$6.95
C-027	Firearms Silencers Volume 1	$13.95
C-029	M1 Carbine Arsenal History	$7.95
C-032	The Silencer Cookbook	$9.95
C-038	How To Build A Beer Can Morter	$4.95
C-040	Criminal Use of False ID	$14.95
C-045	Improvised Weapons of American Underground	$14.95
C-048	Catalog of Military Suppliers	$18.95
C-056	Navy Seal Manual	$29.95
C-065	Poor Man's James Bond Vol 1	$34.95
C-113	Submachine Gun Designers Handbook	$19.95
C-114	AR-15, M16 and M16A1 5.56mm Rifles	$13.95
C-117	Ruger Carbine Cookbook	$13.95
C-120	Shootout II	$14.95
C-122	AK-47 Assault Rifle	$15.95
C-123	Combat Loads for Sniper Rifles	$18.95
C-125	Browning Hi-Power Pistols	$13.95
C-126	UZI Submachine Gun	$13.95
C-127	FullAuto Vol 1 Ar-15 Modification Manual	$11.95
C-128	Full Auto Vol 2 Uzi Mod Manual	$11.95
C-130	Full Auto Vol 4 Thompson SMG	$11.95
C-131	FullAuto Vol 5 M1 Carbine to M2	$11.95
C-135	M-14 Rifle, The	$12.95
C-136	Fighting Garand Manual	$13.95
C-137	Ranger Training & Operations	$19.95
C-138	Emergency War Surgery	$29.95
C-140	Emergency Medical Care/Disaster	$14.95
C-143	Survival Medicine	$15.95
C-150	Improvised Munitions Black Book Vol 1	$17.95
C-151	Improvised Munitions Black Book Vol 2	$17.95
C-152	Impro. Munitions Black Book Vol 3	$29.95
C-155	Fighting Back on the Job	$14.95
C-162	Colt .45 Auto Pistol Manual	$13.95
C-163	Survival Evasion & Escape	$19.95
C-165	USMC Sniping	$19.95
C-175	Beat the Box	$7.95
C-197	How to Make Disposable Silencers Vol. 1	$19.95
C-198	Expedient Hand Grenades	$21.95
C-199	U. S. Army Counterterrorism Training Manual	$19.95
C-203	Explosives and Propellants	$19.95
C-204	Improvised Munitions/Ammonium Nitrate	$19.95
C-205	Two Component High Explosives Mixtures	$19.95
C-207	CIA Improvised Sabotage Devices	$19.95
C-208	Spcial Forces Demolitions Training Handbook	$19.95
C-210	Guerilla Warfare	$17.95
C-211	FN-FAL Auto Rifles	$21.95
C-218	Professional Homemade Cherry Bomb	$15.95
C-219	Improvised Shaped Charges	$19.95
C-220	Improvised Explosives, Use In Detonation Devices	$19.95
C-227	Springfield '03 Rifle Manual	$15.95
C-229	Lock Picking Simplified	$11.95
C-230	How to Fit Keys by Impressioning	$9.95
C-231	Lockout -Techniques of Forced Entr	$15.95
C-235	How to Open Handcuffs Without Keys	$13.95
C-244	Camouflage	$14.95
C-274	M14 and M14A1Rifles and Rifle Marksmanship	$23.95
C-284	Thompson Submachine Guns	$19.95
C-289	M1 Carbine Owners Manual	$13.95
C-290	Clandestine Ops Man/Central America	$14.95
C-324	How to Make Disposable Silencers Vol. 2	$19.95
C-337	Sten MKII SMG Construction Manual	$24.95
C-356	Map Reading and Land Navigation	$19.95
C-365	Survival Gunsmithing	$14.95
C-372	Poor Man's James Bond Vol 2	$34.95
C-386	Self-Defense Requires No Apology	$15.95
C-414	Select Fire 10/22	$15.95
C-415	Construction Secret Hiding Places	$15.95
C-416	Brown's Alcohol Motor Fuel Cookbook	$24.95
C-417	Canteen Cup Cookery	$9.95
C-418	Improvised Batteries/Detonation Devices	$19.95
C-421	Walther P-38 Pistol Manual	$13.95
C-422	P-08 Parabellum Luger Auto Pistol	$14.95
C-423	Sten Submachine Gun, The	$11.95
C-427	How to Train a Guard Dog	$16.95
C-429	The Anarchist Handbook Vol. 1	$14.95
C-430	Beretta - 9MM M9	$15.95
C-432	Mercenary Operations Manual	$14.95
C-439	Improvised Lock Picks	$17.95
C-443	How to Disappear Completely & Never be Found	$19.95
C-455	Modern Day Ninjutsu	$17.95
C-512	FullAuto Vol 8 M14A1 & Mini 14	$11.95
C-518	L.A.W. Rocket System	$14.95
C-528	With British Snipers, To the Reich	$29.95
C-539	Dental Specialist	$59.95
C-558	Food, Fur and Survival	$17.95
C-561	The Squeaky Wheel	$17.95
C-562	Survival Chemist Book	$17.95
C-564	Medical Specialist	$59.95
C-571	U. S. Marine Corps Scout/Sniper Training Manual	$27.95
C-610	Aunt Bessie's Wood Stove Cookbook	$9.95
C-622	Defensive Shotgun	$14.95
C-625	MAC-10 Cookbook	$14.95
C-627	Brown's Lawsuit Cookbook	$26.95
C-628	Can You Survive	$16.95
C-633	Hand to Hand Combat	$11.95
C-634	USMC Hand to Hand Combat	$9.95
C-635	Hand to Hand Combat by D'Eliscue	$11.95
C-636	Prisons Bloody Iron	$16.95
C-639	Police Karate	$17.95
C-646	Science of Revolutionary Warfare	$16.95
C-668	Survival Shooting for Women	$14.95
C-680	Infantry Scouting, Patrol. & Sniping	$16.95
C-694	CIA Field Expedient Incendiary Manual	$19.95
C-761	How to Lose Your X-Wife Forever	$24.95
C-763	Vigilante Handbook	$19.95
C-775	Alcohol Distillers Handbook	$27.95
C-776	How to Build a Junkyard Still	$15.95
C-777	Engineer Explosives of WWI	$19.95
C-794	US Marine Corps Essential Subjects	$19.95
C-815	US Marine Bayonet Training	$11.95
C-818	The Poisoner's Handbook	$29.95
C-823	Cold Weather Survival	$14.95
C-829	The Anachist Handbook Vol. 2	$14.95
C-840	Freedom Road	$21.95
C-889	Napoleon's Maxims of War	$11.95
C-890	Company Officers HB of Ger. Army	$19.95
C-891	German Infantry Weapons Vol 1	$22.95
C-896	Op. Man. 7.62mm M24 Sniper Weapon	$13.95
C-899	US Army Bayonet Training	$11.95
C-9002	German MG-34 Machinegun Manual	$19.95
C-9012	Firearm Silencers Vol 3	$18.95
C-9013	HK Assault Rifle Systems	$27.95
C-9047	SKS Type of Carbines, The	$17.95
C-9048	Rough Riders, The	$24.95
C-9052	Private Weaponeer, The	$17.95
C-9057	Keys To Understanding Tubular Locks	$11.95
C-9060	The Anarchist Handbook Vol. 3	$14.95
C-9080	Leadership Hanbook of Small Unit Ops	$14.95
C-9081	Dirty Fighting	$13.95
C-9082	Lasers & Night Vision Devices	$29.95
C-9083	Water Survival Training	$8.95
C-9084	Assorted Nasties	$29.95
C-9085	Ruger P-85 Family of Handguns	$19.95
C-9101	Guide to Germ Warfare	$17.95
C-9102	CIA Field Expedient Method for Explosives Preparation	$19.95
C-9119	Surviving Global Slavery	$16.95
C-9120	Military Knife Fighting	$14.95
C-9127	H&R Reising Submachine Gun Manual	$19.95
C-9131	Live to Spend It	$29.95
C-9134	Military Ground Rappelling Techniques	$15.95
C-9135	Smith & Wesson Autos	$19.95
C-9137	Caching Techniques of U.S. Army Special Forces	$14.95
C-9138	USMC Battle Skills Training Manual	$49.95
C-9139	Survival Armory	$27.95
C-9156	Concealed Carry Made Easy	$17.95
C-9164	The L'il M-1, The .30 Cal. M-1 Carbine	$17.95
C-9170	Urban Combat	$27.95
C-9178	Apocalypse Tomorrow	$17.95
C-9198	MP40 Machinegun Manual	$16.95
C-9199	Clear Your Record & Own a Gun	$24.95
C-9200	Sig's Handguns Manual	$19.95
C-9210	Sniper Training	$24.95
C-9212	Poor Man's Sniper Rifle	$19.95
C-9221	The Official Makarov Pistol Manual	$14.95
C-9224	The Butane Lighter Hand Grenade	$14.95
C-9228	Unarmed Against the Knife	$14.95
C-9229	Black Book of Booby Traps	$29.95
C-9239	Militia Battle Manual	$22.95
C-9247	Black Book of Revenge	$15.95
C-9253	Bankruptcy, Credit and You	$18.95
C-9255	The Sicilian Blade	$15.95
C-9259	How to Build Practical Firearm Silencers	$18.95
C-9262	Glock's Handguns	$19.95
C-9263	Heckler and Koch's Handguns	$19.95
C-9264	The Poor Man's Ray Gun	$11.95
C-9265	The Poor Man's R. P. G.	$24.95
C-9269	David's Tool Kit	$27.95
C-9279	Credit Improvement & Protection Handbook	$19.95
C-9281	Build Your Own AR-15	$19.95
C-9282	How to Live Safely in a Dangerous World	$19.95
C-9293	Bllack Book of Arson	$24.95
C-9322	How to Build Flash/Stun Grenades	$19.95
C-9342	Entrapment	$19.95
C-9359	The FN-FAL Rifle Et Al.	$21.95
C-9367	The Mental Edge, Revised	$17.95
C-9373	Survival Bible by Duncan Long	$59.95
C-9405	The Internet Predator	$17.95
C-9406	The Criminal Defendants Bible	$49.95
C-9430	Protect Your Assets	$24.95
C-9432	How To Build Military Grade Suppressors	$21.95
C-9441	Mossberg Shotguns	$24.95
C-9478	Poor Man's Primer Manual, The	$24.95
C-9546	Agents HB of Black Bag Ops	$14.95
C-9556	Improvised Rocket Motors	$19.95
C-9612	Homestead Carpentry	$14.95
C-9633	Getaway	$13.95

PRICES SUBJECT TO CHANGE WITHOUT NOTICE

870-862-3811

Desert Publications

215 S. Washington Ave Dept. - BK - 163

El Dorado, AR 71730 USA

Shipping & Handling
1 item $7.95 - 2 or more $12.95